SARDAR SAROVAR PROJECT
ON
THE RIVER NARMADA

History of Design, Planning and Appraisal

Volume 1

About the Editors

R. Parthasarathy is a Professor, Faculty of Planning and Public Policy at CEPT University, Ahmedabad. He has about 25 years of teaching and research experience and has been a Visiting Scholar at the University of California, Berkeley; Professor and later Director of Gujarat Institute of Development Research, Ahmedabad. He has been a member of many government committees and has also been a consultant to DFID, ADB and other national and international organisations. He also works closely with NGOs. His research interests include, water management, large-scale irrigation systems, natural resources, infrastructure management and environmental economics. He has published 3 books and over 70 articles.

Ravindra H. Dholakia has been a Professor of Economics at IIM, Ahmedabad since 1985. He has about 34 years of experience in teaching Economics to different groups like students, executives, policy makers and senior government officials. He has practical experience in high-powered policy making and evaluation bodies in both state and central governments such as the Sixth Central Pay Commission in India (2006-8), State Public Finance Reforms Committee (1998-2001), Expert Committee to Restructure Public Sector Units (2004-7) and Committee to Restructure Public Debt (2003-5), appointed by the Government of Gujarat. Currently, he is a member of the High Level Expert Committee appointed by the Government of India to suggest measures for efficient management of public expenditure (2010-11). He has published 39 monographs, 17 books, and about 130 research papers and international cases on economic planning and policy reforms.

Sardar Sarovar Project on the River Narmada

History of Design, Planning and Appraisal

Volume 1

Editors
R. Parthasarathy
Ravindra H. Dholakia

Published for
CEPT University Press, Ahmedabad-380009
by
Concept Publishing Company Pvt. Ltd., New Delhi-110059

ISBN-13: 978-81-8069-795-1 (Set)
978-81-8069-796-8 (Vol. 1)
978-81-8069-797-5 (Vol. 2)
978-81-8069-798-2 (Vol. 3)

First Published 2011

CEPT UNIVERSITY
Kasturbhai Lalbhai Campus
University Road,
Ahmedabad-380009 (Gujarat) India.
Phone : 26302470, 26302740,
Fax : +91-079-26302075
Email: publications@cept.ac.in
Website: www.cept.ac.in

Printed and Published by

Concept Publishing Company Pvt. Ltd.
Regd. Office:
A/15-16, Commercial Block, Mohan Garden
New Delhi-110059 (India)
Phones : 25351460, 25351794, *Fax* : 091-11-25357109
Email : publishing@conceptpub.com
Website: www.conceptpub.com

Editorial Office:
H-13, Bali Nagar, New Delhi-110 015, India.

Cataloging in Publication Data--*Courtesy:* D.K. Agencies (P) Ltd. <docinfo@dkagencies.com>

Sardar Sarovar Project on the river Narmada / editors, R. Parthasarathy, Ravindra H. Dholakia.
3 v.. cm.
Includes bibliographical references (p.) and index.
Contents: v. 1. History of design, planning and appraisal -- v. 2. History of rehabilitation and implementation -- v. 3. Impacts so far and ways forward.
ISBN 13: 9788180697951 (set) ISBN 13: 9788180697975 (v. 2)
ISBN 13: 9788180697968 (v. 1) ISBN 13: 9788180697982 (v. 3)

1. Sardar Sarovar (Narmada) Project--Congresses. 2. Water resources development--Environmental aspects--India--Maharashtra--Congresses. 3. Water resources development--Environmental aspects--India--Gujarat--Congresses. 4. Dams--India--Maharashtra--Design and construction--Congresses. 5. Dams--India--Gujarat--Design and construction --Congresses. 6. Rehabilitation--India--Maharashtra--Congresses. 7. Rehabilitation--India--Gujarat--Congresses. 8. Irrigation--India--Maharashtra--Congresses. 9. Irrigation--India--Gujarat--Congresses. I. Parthasarathy, R., Dr. II. Dholakia, Ravindra H., 1953- III. Center for Environmental Planning and Technology University (Ahmadābād, India)

DDC 333.910095479 22

Sardar Sarovar Project on the River Narmada

Impacts So Far and Ways Forward

Volume 3

Editors

R. Parthasarathy

Ravindra H. Dholakia

Published for

CEPT University Press, Ahmedabad-380009

by

Concept Publishing Company Pvt. Ltd., New Delhi-110059

ISBN-13: 978-81-8069-800-2 (Set)
978-81-8069-801-9 (Vol. 1)
978-81-8069-802-6 (Vol. 2)
978-81-8069-803-3 (Vol. 3)

First Published 2011

CEPT UNIVERSITY
Kasturbhai Lalbhai Campus
University Road,
Ahmedabad-380009 (Gujarat) India.
Phone : 26302470, 26302740,
Fax : +91-079-26302075
Email: publications@cept.ac.in
Website: www.cept.ac.in

Printed and Published by

Since 1974

Concept Publishing Company Pvt. Ltd.
Regd. Office:
A/15-16, Commercial Block, Mohan Garden
New Delhi-110059 (India)
Phones : 25351460, 25351794, *Fax* : 091-11-25357109
Email : publishing@conceptpub.com
Website: www.conceptpub.com

Editorial Office:
H-13, Bali Nagar, New Delhi-110 015, India.

Cataloging in Publication Data--*Courtesy:* D.K. Agencies (P) Ltd. <docinfo@dkagencies.com>

Sardar Sarovar Project on the river Narmada / editors, R. Parthasarathy, Ravindra H. Dholakia.
3 v.. cm.
Includes bibliographical references (p.) and index.
Contents: v. 1. History of design, planning and appraisal -- v. 2. History of rehabilitation and implementation -- v. 3. Impacts so far and ways forward.
ISBN 13: 9788180698002 (set) ISBN 13: 9788180698026 (v. 2)
ISBN 13: 9788180698019 (v. 1) ISBN 13: 9788180698033 (v. 3)

1. Sardar Sarovar (Narmada) Project--Congresses. 2. Water resources development--Environmental aspects--India--Maharashtra--Congresses. 3. Water resources development--Environmental aspects--India--Gujarat--Congresses. 4. Dams--India--Maharashtra--Design and construction--Congresses. 5. Dams--India--Gujarat--Design and construction --Congresses. 6. Rehabilitation--India--Maharashtra--Congresses. 7. Rehabilitation--India--Gujarat--Congresses. 8. Irrigation--India--Maharashtra--Congresses. 9. Irrigation--India--Gujarat--Congresses. I. Parthasarathy, R., Dr. II. Dholakia, Ravindra H., 1953- III. Center for Environmental Planning and Technology University (Ahmadābād, India)

DDC 333.910095479 22

Shri Narendra Modi
Chief Minister, Gujarat

Foundation stone of the Sardar Sarovar Dam on the river Narmada was laid on 5th April 1961 hence this Project will also be completing its long journey of five decades as Gujarat State is celebrating its Golden Jubilee. Sardar Sarovar Project is not only an icon of water resources development but it is also an expression of the strong will of the people of this State. Against a portrayed perception of SSP as a controversial project, this compilation reveals its multi-faceted dimensions. Not only the dimensions and benefits of this project are awesome but it has also got many innovative features in planning, designing, implementation, financing and managing. It is one of the most studied water resources projects in the world and its implementation has been a long struggle against all odds, which could be possible mainly due to wide public support and participation from people of Gujarat. Successive Governments made positive contribution in this effort, and the will of the people of Gujarat has been unstinting, and therefore in a true sense SSP is Project of the People of Gujarat.

Last one decade has seen accelerated development in this Project as a result of which the dam height could be raised to the spillway crest level, two hydro power houses could be made operational, 458 kilometer long Narmada Main Canal could be completed and command area of more than 5 lakh hectare could be developed by constructing a canal network of 19,000 kilometer length. Increased and assured availability of Narmada water for irrigation has helped in achieving and sustaining an impressive agricultural growth rate of more than 9% in the State against the national average of about 3% during the same period.

SSP is no less than a *"living technological university"* and the need for proper documentation of its evolution and growth was increasingly felt. As one of the firsts, these three volumes have now documented history of many facets of this very important project and have captured the highlights of this unique endeavor, while also suggesting ways forward.

The articles written by pioneer engineers, involved activists, academicians and government officials present various aspects of the project and Editors too have done a commendable job in bringing the individual contributions to a comprehensive publication level. Articles in this volumes address several concerns raised by critics at various stages of the Project. In my opinion on this 'World Water Day' during the golden jubilee year of the State, there can not be a better tribute to the people of Gujarat who have wholeheartedly supported the Sardar Sarovar Project in this journey. Sardar Sarovar Narmada Nigam Limited also deserves compliments for having supported this invaluable commemorative effort that will help the future generations to understand the evolution and growth of Sardar Sarovar Project.

(Narendra Modi)

Preface

The Sardar Sarovar Project (SSP) on the River Narmada has a long and interesting, yet sometimes controversial history. Even before the formal proposal was made in 1959 for a dam at the Sardar Sarovar site at Navagam, there had been various attempts to harness the Narmada. While Independence solved the problems of political complexity associated with organisation among the princely states for construction of the dam, the reorganisation of states on linguistic basis in 1956 added a new dimension. When preliminary work on the dam began in 1961, the disagreements between the riparian states on the sharing of waters led to the project being suspended. In 1969 the Narmada Water Disputes Tribunal (NWDT) was set up to decide the inter-state allocation of water and the costs of the dams and other infrastructure needed to exploit the river. In 1979 the Tribunal Award was made and full-scale construction of Sardar Sarovar began in 1987.

In its long history, SSP has faced many hurdles. It was not until the year 2003-04 that many of the issues relating to construction of the dam were resolved. Meanwhile, the main canal was constructed and many minors and sub-minors are in the process of being built. Other important features of this project are the rehabilitation of the project affected persons and the planning of use of its waters. While many of the controversies have been well documented, many salient features associated with the project have not been. It is precisely the many dimensions of the setting up of this project that acted as the stimulus to document the evolution and realisation of the project during the Golden Jubilee year of the state of Gujarat. To this end, three main themes emerged, namely, (a) the History of Design,

Planning and Appraisal including the Environment; (b) the Resettlement and Rehabilitation Process during Implementation; and (c) the Future Outlook and Way Forward.

The idea of producing edited volumes on these themes was discussed in a half-day Workshop at CEPT University on August 6, 2010. The Workshop, titled 'Evolution and Growth of Sardar Sarovar: The Life Line of Gujarat was attended by Engineers, most of whom have superannuated, but who had been involved in the design and construction of the dam and the canals; academics, NGO activists, serving and retired officials of Sardar Sarovar Narmada Nigam Limited (SSNNL); practitioners and policy researchers. In many ways this was an exploratory workshop; some came with the notion of processes being important; some focused on the history, and yet others suggested that the many innovations during the project reflect the flexible and sensitive responses to the changing environments and challenges moving away from the conventional designs. Continued discussions with the participants and the editors evolved into a commitment to contribute to the volumes well-researched articles on the different themes. It was also resolved that this volume need not go into the many controversies but should focus on processes and innovative aspects of SSP.

Thus the important feature of these volumes is not only that most of its contributors were directly involved in the project at one time or the other, but also had intimate knowledge of the various interrelated issues. As discussed in the August Workshop, the contributed papers before being published were to go through varied and multiple levels of review. The draft papers were discussed at CEPT University in the Second Workshop held from October 29 to 31, 2010. During this Workshop special invited lectures by Professor Y.K. Alagh (former Union Minister of Planning and Programme Implementation, Science and Technology and Power, GoI, and, Vice-Chairman of the Narmada Planning Group); Dr. C.C. Patel (former Secretary, Ministry of Water Resources, GoI, and formerly CMD of SSNL); Mr. Jay Narayan Vyas, Minister of Health and Tourism, and formerly Minister of Narmada Development, GoG); and Mr. I.P. Gautam (Municipal Commissioner, Ahmedabad and formerly, Director Finance, SSNNL) were also arranged. In fact, we later persuaded Mr. Jay Narayan Vyas and Mr. I.P. Gautam to convert their lectures into papers for the volumes.

During the Workshop the peer discussions on the papers contributed immensely to further sharpening the focus. Some of the authors also presented the final version of their papers during the two days of January 3 and 4, 2011 in the final workshop held at CEPT University.

An effort of this magnitude would not have been possible but for the strong and committed response of the authors. The papers were subjected to a refereeing process and some of them were re-reviewed before acceptance as publishable material. For this publication, we also had an Advisory Board consisting of Mr. N.B. Desai, Mr. N. Ramaswami, Mr. G.L. Java, Mr. P.H. Rana, Dr. Anil Patel, Mr. Hasmukh Patel, Dr. Sanat Chavan, Dr. Tushaar Shah and Dr. M.B. Joshi. We are indeed thankful to the authors, the referees and the Advisory Committee Members for their invaluable support in bringing these volumes within the time constraints. Our special thanks go to Mr. Vivek Khadpekar for his commitment and involvement with the editors while carrying out a commendable and a very competent editing of the papers.

All the analysis, views and opinions contained in the papers are those of the respective authors and do not reflect the views of the organisations they belong. Similarly, while we have taken all possible care to check the facts and figures presented in the paper, the errors of omission and commission still rest with the authors of the papers.

We are very grateful to Mr. S. Jagadeesan, Mr. K. Srinivas and Mr. G.R. Aloria of SSNNL, Gujarat for generously sponsoring and participating in the various workshops during this year of Swarnim Gujarat celebration. The present volumes are the outcome of these workshops. Dr. M.B. Joshi, along with his colleagues at the SSNNL, Dr. K.D. Acharya and Dr. Vinod Rana, provided very crucial support at all stages during the work including providing data, documents, maps and many reference materials. Ms. Mahima Gupta, an irrigation engineer and a planner herself, provided able research support; Ms. Aparna, Mr. Harish Joshi and Ms. Soumini Raja were always willing to help us not only during the organisation of the three workshops but also during the various stages of bringing out these volumes.

At the CEPT University, Dr. R.N. Vakil was always very encouraging right from the time the plan to bring out these volumes was discussed. In fact he speeded up the setting up of the CEPT University Press and the Centre for Documentation and Publication at the University to facilitate the work. Ms. Nirmala Khadpekar, Associate Professor at CEPT University took charge of the publication of these volumes and provided able and unflinching support. Professors Utpal Sharma and Shivanand Swamy not only provided crucial support but, along with Professor Ashwani Kumar, also attended some of the workshops and gave valuable feedback. We are also thankful to the Faculty members for their encouragement. A work of this kind requires administrative and library support. The CEPT University administration, particularly the Registrar, Staff at Centre for Research, Development and Consultancy, the library staff and Jitubhai and Dhuljibhai provided this critical support. We deeply appreciate and thank all of them.

The management team of the Concept Publishing Company was extremely cooperative and efficient in bringing out these volumes in time.

Ahmedabad
March 2011

R. Parthasarathy
Ravindra H. Dholakia

Contents

Part 2: Planning for SSP

Part 3: Appraisal for SSP

Volume 2: History of Rehabilitation and Implementation

Part 4: Resettlement and Rehabilitation in SSP

Part 7: Ways Forward

Abbreviations

AIBP	Accelerated Irrigation Benefit Programme
AISLUS	All India Soil and Land User Survey
AMC	Ahmedabad Municipal Corporation
APMC	Agricultural Produce Marketing Cooperative
ARCH	Action Research in Community Health and Development
ARWSP	Accelerated Rural Water Supply Programme
ASS	Argillaceous Sand Stone
BARC	Bhabha Atomic Research Center
BCM	Billion Cubic Metres
BHEL	Bharat Heavy Electricals Limited
BIP	Broach Irrigation Project
BIS	Bureau of Indian Standards
BMC	Bombay Municipal Corporation
BOC	Board of Consultants
BOT	Build-Operate-Transfer
CA	Compensatory Afforestation
CAD	Catchment Area Development
CAPs	Canal-Affected-Persons
CARE	Credit Analysis and Research Ltd
CAT	Catchment Area Treatment
CBR	Costs-Benefit Ratio
CCA	Cultural Command Area
CDO	Central Designs Organisation, Gujarat
CEO	Chief Executive officer
CES	Consulting Engineering Services
CGWB	Central Ground Water Board
Ch.	Chainage
CHPH	Canal Head Power House
CI	Confidence Interval

CICFRI	Central Inland Capture Fisheries Institute
CMCE	Critical Maximum Credible Earthquake
CRISIL	Credit Rating and Information Services of India Ltd
CSS	Centre for Social Studies
CSWCR&TI	Central Soil and Water Conservation Research and Training Institute
Cumecs	Cubic Meter Per second
CVC	Constant Volume Canals
CW&PC	Central Water & Power Commission
CWC	Central Water Commission
CWINC	Central Waterways, Irrigation and Navigation Commission
CWPC	Central Water and Power Commission
CWPRS	Central Water and Power Research Station
D&R	Design and Research
DACNET	Directorate of Economics and Statistics
DAP	Diammonium Phosphate
DBE	Design Basis Earthquake
DCC	Data Collection Centre
DDB	Deep Discount Bonds
DDO	District Development Officer
DDRP	Dam Design Review Panel
DF	Design Floods
DMAs	District Metering Areas
DOC	Divisional Operation Centres
DoEF	Department of Environment and Forest
DSC	Development Support Centre
DSP	Dam Safety Panel
DST	Department of Science and Technology
EDF	Environment Defense Fund
EIA	Environmental Impact Assessment
EPC	Engineering, Procurement and Construction
EPGA	Effective Peak Ground Acceleration
ERR	Economic Rate of Return
ESG	Environment Subgroup
FAO	Food and Agriculture Organization
FCA	Forest Conservation Act
FCI	Food Corporation of India
FEM	Finite Element Material
FFDA	Fish Farmers Development Authority
FMG	Five Member Group

FRL	Full Reservoir Level
FSD	Full Supply Depth
FSL	Full Supply Level
GAD	General Administration Department
GCA	Gross Cropped Area
GCEL	Gujarat Communication &. Electrical Ltd
GCMMF	Gujarat Cooperative Milk Marketing Federation
GDP	Gross Domestic Product
GEB	Gujarat Electricity Board
GEER	Gujarat Ecological Education and Research Foundation
GERI	Gujarat Engineering Research Institute
GGRC	Gujarat Green Revolution Company
GIS	Geographical Information System
GNFC	Gujarat Narmada Fertilizer Company
GoG	Government of Gujarat
GoI	Government of India
GoM	Government of Maharashtra
GoMP	Government of Madhya Pradesh
GR	Government Resolution
GRA	Grievances Redressal Authority
GRC	Grievances Redressal Cell
GSDP	Gross State Domestic product
GSFC	Gujarat State Fertilizer Company
GSI	Geological Survey of India
GWIL	Gujarat Water Infrastructure Limited
GWRDC	Gujarat Water Resources Development Corporation
GWSSB	Gujarat Water Supply and Sewerage Board
Ha	Hectare
HB	Hybrid
HH	Household
HP	Hand Pump
HYS	High Yield Strength
HYV	High-Yielding Variety
IAS	Indian Administrative Services
IBM-IFSSR	Inter-State Board for Management of Integrated Fisheries in the Sardar Sarovar Reservoir
IBPT	Irrigation By Pass Tunnel
ICAR	Indian Council of Agricultural Research
ICICI	Industrial Credit and Investment Corporation of India
ICID	International Congress on Irrigation and Drainage

ICOLD	International Commission on Large Dams
ICOR	Incremental Capital-Output Ratio
ICS	Indian Civil Services
ICSSR	Indian Council of Social Science Research
ID	Irrigation Department
IDA	International Development Association
IDBI	Industrial Development Bank of India
IFPRI	International Food Policy Research Institute
IIM	Indian Institute of Management
IITM	Indian Institute of Tropical Meteorology
ILO	International Labour Organization
IMD	India Meteorological Department
IMT	Irrigation Management Transfer
INR	Indian Rupee
IPCC	Intergovernmental Panel on climate Change
IR	Independent Review
IRM	Independent Review Mission
IRMA	Institute of Rural Management Ananad
IRR	Internal Rate of Return
IS	Indira Sagar
ISP	Indira Sagar Project
ISRO	Indian Space Research Organization
IWMI	International Water Management Institute
J&K	Jammu and Kashmir
JNNURM	Jawaharlal Nehru National Urban Renewal Mission
JP	Jayaprakash Narayan
KBC	Kachchh Branch Canal
LA	Loan Agreement
LF	Load Factor
LIC	Life Insurance Corporation
LPC	Land Purchase Committee
LPCD	Litres Per Capita per Day
LVFO	Lake Victoria Fisheries Organization
M&I	Municipal and Industrial
MAF	Million Acre-Feet
MC	Municipal Corporation
MCC	Main Control Centre
MCE	Maximum Credible Earthquake
MCM	Million Cubic Meter
MDDL	Mean Draw Down Level

MDE	Maximum Design Earthquake
MEQ	Micro Earth Quake
MHa	Million Hectares
MHM	Million Hectare Meter
MHUPA	Ministry of Housing & Urban Poverty Alleviation
MIS	Management Information System
MLD	Million Litres per Day
MNP	Minimum Needs Programme
MOEF	Ministry of Environment and Forests
MOUD	Ministry of Urban Development
MOWR	Ministry of Water Resources
MP	Madhya Pradesh
MPISSR	Madhya Pradesh Institute of Social science Research
MSP	Minimum Support Price
MSU	Maharaja Sayajirao University
MW	Megawatt
MWL	Maximum Water Level
NABARD	National Bank for Agriculture and Rural Development
NBA	Narmada Bachao Andolan
NCA	Narmada Control Authority
NCSDP	National Committee on Seismic Design Parameters
NDDB	National Dairy Development Board
NDS	Narmada Dharangrast Samiti
NEERI	National Environmental Engineering Research Institute
NFHS	National Family Health Survey
NGNS	Narmada Ghati Navnirman Samiti
NGO	Non-Government Organisation
NGRI	National Geophysical Research Institute
NHPC	Narmada High Power Committee
NIFDA	Narmada Inter-State Fisheries Development Authority
NIH	National Institute of Hydrology
NMC	Narmada Main Canal
NOF	Non-Overflow
NPDDC	Narmada Project Dam Designs Circle
NPG	Narmada Planning Group
NPK	Nitrogen, Phosphorus and Potassium
NPSSM	Narmada Project Steady State Model
NPUSM	Narmada Project Unsteady State Model (NPUSM)
NRSA	National Remote Sensing Agency
NRSC	National Remote Sensing Centre

NRW	Non-Revenue Water
NSA	Net sown area
NSP	Narmada Sagar Project
NSSO	National Sample Survey Organisation
NSV	No Source Village
NTPC	National Thermal Power Corporation
NVDA	Narmada Valley Development Authority
NWDT	Narmada Water Disputes Tribunal
NWDTA	Narmada Water Dispute Tribunal Award
NWRDC	Narmada Water Resources Development Committee
O&M	Operations and Maintenance
OBC	Other Backward Castes
OBE	Operating Basis Earthquake
OECD	Organisation for Economic Cooperation and Development
OECF	Overseas Economic Co-operation Fund
ONGC	Oil and Natural Gas Corporation
ORG	Operation Research group
PAFs	Project-Affected Families
PAPs	Project-Affected Persons
PAS	Performance Assessment System
PCR	Project Completion Report
PF	Provident Fund
PFC	Power Finance Corporation
PGA	Peak Ground Acceleration
PIM	Participatory Irrigation Management
PINS	Pressurised Irrigation Network System
PMF	Probable Maximum Floods
PMP	Probable Maximum Precipitation
PPP	Public Private Partnership
PUCL	People's Union for Civil Liberties
QL	Quintal
R & D	Research and Development
R&E	Research and Evaluation
R&R	Resettlement and Rehabilitation
RBPH	River Bed Power House
RCC	Reinforced Cement Concrete
RCNCA	Review Committee of NCA
RDSO	Railway Designs and Standards Organisation
RIDF	Rural Infrastructure Development Fund
RIS	Reservoir Induced Seismicity

RL	Reduced Level
RO	Reverse Osmosis
RRWHSs	Rooftop Rain Water Harvesting Systems
RTU	Remote Terminal Unit
RWSS	Regional Water Supply Scheme
SAR	Staff Appraisal Report
SBC	Saurashtra Branch Canal
SC	Scheduled castes
SCADA	Supervisory and Control Data Acquisition
SCBA	Social Cost-Benefit Analysis
SCF	Standard Conversion Factor
SD	Standard Deviation
SHGs	Self help Groups
SJMMSVY	Swarna Jayanti Mukhya Mantri Shaheri Vikas Yojana
SOI	Survey of India
SPF	Standard Projected Flood
SPIESR	Sardar Patel Institute of Economic & Social Research
SPPWCP	Sardar Patel Sahbhagi Jal Sanchay Yojana
SPV	Special Purpose Vehicle
SRR	Social Rate of Return
SS	Sardar Sarovar
SSCAC	Sardar Sarovar Construction Advisory Committee
SSD	Sardar Sarovar Dam
SSL	Suspended Sediment Load
SSNNL	Sardar Sarovar Narmada Nigam Limited
SSP	Sardar Sarovar Project
SSPA	Sardar Sarovar Punarvas Agency
SSRA	Sardar Sarovar Rehabilitation Agency
ST	Scheduled Tribes
STPR	Social Time Preference Rate
SWRC	Social Work and Research Centre
TCIL	Telecommunication Consultants India Limited
TDO	Taluka Development Offcier
TDS	Total Dissolved Solids
TECS	Tata Economic Consultancy Services
TISS	Tata Institute of Social Sciences
TMC	Thousand Million Cubic
UIDSSMT	Urban Infrastructure Development Scheme For Small and Medium Towns
UIP	Ultimate Irrigation Potential

ULBs	Urban Local Bodies
UN	United Nations
UNCED	United Nations Conference on Environment and Development
UNDP	United Nations Development Programme
UNICEF	United Nations Children's Fund
UNIDO	United Nations Industrial Development Organisation
UP	Uttar Pradesh
USBR	United States Bureau of Reclamation
USSR	Union of Soviet Socialist Republics
VLW	Village Level Worker
VSA	Village Service Area
WASMO	Water and Sanitation Management Organisation
WB	World Bank
WCD	World Commission on Dams
WHO	World Health Organization
WIHG	Wadia Institute of Himalayan Geology
WUA	Water Users Association
WWF	World Wildlife Fund

List of Contributors

Acharya, K. D., Assistant Engineer, Sardar Sarovar Narmada Nigam Limited, Gandhinagar. acharya_kishore2005@yahoo.co.in

Bassi, Nitin, Institute for Resource Analysis and Policy (IRAP), Hyderabad.

Chandra, A., Former Student, IRMA, Anand.

Chavan, Sanat, Former Principal Chief Conservator of Forests, Gujarat, A-2/3, Kabir Housing Complex, Opp. Don Bosco School Makarepura Road, Vadodara.

Das, Biswaroop, Professor, Centre for Social Studies, University Campus, Udhana Magadalla Road, Surat. dbiswaroop@gmail.com

Datta, K. Samar, Professor, Indian Institute of Management, Ahmedabad. sdutta@iimahd.ernet.in

Desai N.B., Former Secretary (Narmada), Government of Gujarat, 11, Triveni Park, Opp. S.N.D.T. College, Akota Stadium Area, Vadodara.

Desai, Vaidya Prerana, Sneh-Setu Foundation, Vadodara, snehsetufdn@gmail.com

Dholakia, H Ravindra, Professor, Indian Institute of Management, Ahmedabad. rdholkia@iimahd.ernet.in

Garg, Amit, Professor, Indian Institute of Management, Ahmedabad. amitgarg@iimahd.ernet.in

Gautam, I.P., Municipal Commissioner, Ahmedabad and Former Executive Director, Resources, and Director of Finance, SSNNL, ipgautam@egovamc.com

Goswami, Subhrangsu, Assistant Professor, Faculty of Planning and Public Policy, CEPT University, Ahmedabad. s.goswami@cept.ac.in

Gupta, R.L., President, Jaypee Ventures Limited, NOIDA, Uttar Pradesh

Iyengar, Sudarshan, Vice Chancellor, Gujarat Vidyapith, Ahmedabad. sudarshan54@gmail.com

Jagadeesan, S., IAS, Managing Director, Sardar Sarovar Narmada Nigam Limited, Gandhinagar. sjagadeesan53@gmail.com

Java, G.L., Former Director (SSNL) and Secretary (Retired), Narmada Water Resources and Water Supply Department, Government of Gujarat, gl7java@yahoo.com

Jhala, Neha, Research Associate, Indian Institute of Management, Ahmedabad, nehajhala@ iimahd.ernet.in

Joshi, M.B., Superintending Engineer, Sardar Sarovar Narmada Nigam Limited, Gandhinagar. dr.mbjoshi@gmail.com

Kankal, Bhushan, Research Associate, Indian Institute of Management, Ahmedabad, bhushank@ iimahd.ernet.in

Krishnan, Sunderrajan, Director, CAREWATER, INREM Foundation, Anand. carewater@gmail.com

Kumar, M. Dinesh, Executive Director, Institute for Resource Analysis and Policy (IRAP), Hyderabad. dinesh@irapindia.org

Mehta, Ambrish, ARCH Vadodara, E-702, Samrajya Complex, Near Fatehgunj Post Office, Fatehgunj, Vadodara. m_ambrish@hotmail.com

Mehta, Dinesh, Professor Emeritus, Faculty of Planning and Public Policy, CEPT University, Ahmedabad. dineshmehta@cept.ac.in

Mehta, Meera, Professor Emeritus, Faculty of Planning and Public Policy, CEPT University, Ahmedabad. meeramehta@cept.ac.in

Parmar, B.J., Former Secretary, Water Resources Department, Government of Gujarat, bjpsays@rediffmail.com

Parthasarathy, R., Professor, Faculty of Planning and Public Policy, CEPT University, Ahmedabad, r.parthasarathy@yahoo.com

Patel, Anil, Arch Mangrol, Rajpipla, District Narmada

Pathak, Mahesh, Former Chairman, Gujarat State Agriculture Commission and Former Executive Member, Narmada Planning Group, Government of Gujarat, mhpathaks@yahoo.in

Pathak, Swati, Research Associate, Indian Institute of Management, Ahmedabad, swatip@ iimahd.ernet.in

Pullabhotia, Hemant Kumar, Research Analyst, International Food Policy Research Institute, NASC Complex, CG Block, Dev Prakash Shastri Road (Opp. Todapur) Pusa, New Delhi 110012.

Rathod, K.G., Chief Engineer and Director(Canals), Sardar Sarovar Narmada Nigam Ltd, Government of Gujarat, Gandhinagar, dire-canal-ssnnl@gujarat.gov.in

Sah, D.C., Professor & Director, Madhya Pradesh Institute of Social Science Research, 6 Bharatpuri Institutional Area, Ujjain, Madhya Pradesh. dc_sah@yahoo.co.uk

Shah, Tushaar, Senior Fellow, International Water Management Institute, Anand. tushaar@irma.ac.in

Sheth, Pravin, Former Professor and Head of Political Science Department at Gujarat University, Ahmedabad. pravinsheth@hotmail.com

Srinivas, K., IAS, Joint Managing Director, Sardar Sarovar Narmada Nigam Limited, Gandhinagar. ksrinivas_ias@yahoo.co.in

Sudhir, C., Former Student, IRMA, Anand

Tomar, Shubhra S., ICSSR Doctoral Fellow, MPISSR, 6 Bharatpuri Institutional Area, Ujjain, Madhya Pradesh

Verma, Shilp, UNESCO-IHE Institute for Water Education, PO Box 3015, 2601 DA Delft, The Netherlands.

Vyas, Jay Narayan, Minister of Health and Family Welfare, Government of Gujarat and Ex-Chairman, Sardar Sarovar Narmada Nigam Limited, min-health@gujarat.gov.in

20

Sardar Sarovar Project: Performance So Far, Plans and Opportunities

R. Parthasarathy and *Ravindra H. Dholakia*

Performance So Far and Opportunities

India's water sector is depicted as being, on the one hand, of turbulent and muddied with the crisis of governance failure and, on the other hand, the absence of a holistic vision and planning for sustainable resource development and management regimes for the future. While the freshwater stock is shrinking, the demand for it is slated to be high in the face of population growth, urbanisation and the resultant growth in demand for water emerging from non-conventional sectors. Groundwater resources are in a critical state in most parts of the country as they account for the largest source of irrigation water supplies (around 70 per cent) and 80 per cent of the domestic water supplies (World Bank, 2005).

The development of water resources, which assumed the centre-stage of planned development programmes in India, had a definite impact on the agrarian front in terms of achieving self-sufficiency in agricultural production along with direct and indirect benefits emanating from forward and backward linkages. However, the plethora of challenges that water sector is fraught with looms large. Most important of these challenges pertain to: (a) the perceptible gap in the provision of safe drinking water across the rural and urban areas; (b) state failure in appreciating water as a fundamental right;

(c) unbridled growth of water markets even in the rural fringes; (d) non-responsiveness of the public institutional and governance systems towards the globalisation-induced dynamic agrarian changes taking place in the canal commands; (e) limited success of the reform measures aimed at through irrigation management; (f) growing inefficiency in the functioning and performance of irrigation systems; and (g) growing environment and human health-related concerns along with the socio-economic impacts of ill-conceived and poorly implemented rehabilitation and resettlement programmes, to mention a few.

While concerns about the impending water crisis in India are soaring, uncertainties prevail about the type, nature and magnitude of problems confronting the water sector in the country. This is due to the sheer lack of consensus on the approaches and methodologies needed to understand the problems and the policies, governance systems and institutional interventions needed to circumvent them. Arguably, the Constitution of India has provided a solid foundation for evolving any type of legal and policy frames required for the water resources sector in the country, as the Constitution considers water a state subject and 'irrigation' being slated as entry 17 of the State List. Further, 'water rights' irrespective of being well-or ill-defined and/or implemented or non-implemented in whatsoever fashion, are derived from the fundamental rights of the Constitution (under Article 21). As water is included in the State List (List II) of the Constitution of India, State governments are empowered to legislate on water related matters.

In fact the country as a whole, and for that matter the individual states, have been following a centralised approach in respect of water resources development and management ever since Independence. The institutional as well as governance paradigms have also been driven by the centralist decision-making frame characterised by top down approach. However, a transformation in the scenario was sought with the 73rd and 74th Amendments to the Constitution in 1992, which have empowered the Panchayati Raj Institutions (PRIs) and encouraged a decentralised management strategy. Following this, there have been a series of enactments and formulations of laws and policies concerning the water sector (surface water, groundwater and drinking water) with a paradigm shift in perspectives and approaches, mostly oriented to the micro-level contexts of individual states. In a majority of cases, these legislations, enactments and policy formulations may be considered as refinements or modifications or additions to the pre-existing legal and regulatory regimes of colonial vintage. For instance, the pre-existing enactments which were amended were the Bombay Irrigation Act (1879) for the western region, covering Gujarat and Maharashtra; the Northern India Drainage and Canal Act (1873) for the northern region states of Uttar Pradesh, Punjab, Haryana and Rajasthan; the Bengal Irrigation Act (1896) as applied to West Bengal and Bihar; Madras Irrigation Cess Act (1865), Irrigation Tanks Improvement Act (1949) and the Tamil Nadu Panchayats' Act, 1958 for Tamil Nadu state.

Of late, India's water sector is engaged in several experimentations involving legal and policy reforms as well as institutional interventions broadly aimed at securing a sustainable water future for the country. These legal and policy level reforms, especially at the level of the states may be said to have obtained vigour and impetus from the twin policies at the national level—the Water Policies of 1987 (GoI, 1987) and 2002 (GoI, 2002) respectively. Following this, a series of legislative and policy oriented reforms have been introduced by the states, including the Gujarat Water Users' Participatory Irrigation Management Act, 2007.

As we had seen in the previous two volumes, Gujarat has taken many innovative steps in the process of setting up the mammoth SSP. For a state made up largely of arid or semi-arid climes, it is not difficult of imagine that careful planning would have gone into utilising water resources. The document *Planning for Prosperity* (SSNNL, 1989) outlines the plan of water distribution. In fact, when the plans are implemented, SSP would be the irrigation system that truly takes cognisance of water management, going beyond merely distribution. In fact, even in the existing irrigation schemes there has been considerable thought given to improving irrigation efficiencies, which are abysmally low not only in Gujarat but in other states in India too (Parthasarathy, 2010). This call for further modernising the way waters, particularly the way large scale irrigation systems is being managed. Enhanced analysis and design of private investments coupled with defined cost-sharing rules could be one of the ways forward to the much discussed irrigation system maintenance issues.

These improvements can possibly then align incentives and make the commitments for operationalising SSP canals, distributory, minors and field channels more credible. They would also help shift government investments towards progressive improvement that aids adaptive problem-solving in irrigation co-management. This is the need of the hour since more than half of the canal systems are yet to be completed. In fact, it offers an opportunity to put in place new practices and investment policies.

While irrigation is an extremely important and productive use of water in a water scarce state, other uses of water—in industries, households and animal husbandry—are equally important. Water also has great relevance in bringing about social change. Therefore, it is a critical resource as well as a product significantly augmented by SSP for the country as a whole and particularly for the four states of Gujarat, Madhya Pradesh, Maharashtra and Rajasthan. The award of the Tribunal specifying allocation of water among these four states resolved the interstate dispute in its use. What is now critical is the allocation of SSP water among various competing sectors within the states. The principle followed for this purpose has to reflect the relative utility and scarcity values of water to the respective societies of the states. If we assume rationality in socially optimising allocation of SSP water among sectors and states by the award, it has an interesting implication. In practice, even if the available water allocation is marginally tampered with in the short-run, it would not lead to any reduction in overall social welfare (Dholakia: Volume 1 in the present series, Chapter 10). In Gujarat, the drinking water needs of several municipalities, villages and habitations are met with SSP water. Similarly, since progress on the canal network as well as construction of the dam to full height has been frequently interrupted, the use of SSP water in practice had to be marginally different from what the Award has given earlier.

Overview: Plans and Opportunities

In Chapter 21 N.B. Desai focuses on the aspects of irrigated agriculture due to SSP and describes the way the command area was planned for the purpose of irrigation. He also dwells on plans such as canal automation, micro-irrigation, strategies for conjunctive use of water and the framework for vertical and horizontal drainage to tackle the problem of waterlogging. The canal automation proposed to be introduced in a phased manner is expected to provide quick, reliable and equitable response to water demands of even the tail reaches and to avoid wastage of precious water. The physical efficiency of water use and thereby its economic efficiency were envisioned to manage the demand for water in agriculture. He also points out that by using drip irrigation, the net saving of water can be as high as 40-50 per cent for most of the crops. Research has also shown that adoption of drip system results not only in water saving but also increases yields and reduces input costs. Further, micro-tube drip and sprinkler technologies require low capital investments and are easy to operate and maintain. Both farmers with small land holdings and large holdings can use this technology to irrigate vegetables, cotton, castor and horticulture crops.

SSP has recognised the importance of drainage systems and has planned and implemented an integrated framework of vertical as well as horizontal drainage, each fulfilling its technical and environmental role. The project also prepared a manual for surface drainage systems in 1994 through experts in the field and a pilot study on the command area of phase I is to be taken up. SSP has integrated specific strategies for conjunctive use of water in the command area. The farmers and farmer organisations are expected to develop new and additional groundwater in the Village Service Areas (VSAs) through tubewells, borewells and open wells because of limited surface water allocations. The importance of education and training of farmers has been clearly recognised by the project. The Water and Land Management Institute and its three regional centres are expected to train farmers, besides irrigation and agricultural engineers.

Field evaluation of the impacts of the project on agriculture has shown that there are changes in the cropping pattern irrespective of the size of land holding. Encouraging conditions have been created for optimum cropping pattern and higher cropping intensity. This has led to the rise and stability in the income of farmers. Small and marginal farmers have started availing of the benefits of irrigation. The author points out that the issues like overestimation of water availability, unrealistic estimation of irrigation efficiency, groundwater and conjunctive use which were raised at the time of construction have been proved to exaggerated. He also states that some concerns which were expressed are unfounded, while other needs to be resolved.

Amit Garg *et al.* in Chapter 22 present the impacts of Narmada water on agriculture in three districts of Gujarat—Ahmedabad, Gandhinagar, and Vadodara—which started receiving the canal water from 2005 onwards. The key question of the study is whether the timely availability of the Narmada water had caused an increase in crop yield and net area sown (NSA). The results show that the Narmada water has significantly increased the average crop yields and NSA of wheat, oilseeds and pulses in all the three districts. In the aggregate, the farmers have also benefited economically, the authors claim. The chapter is organised in six sections. A brief background of Sardar Sarovar Canal in section one, followed by a section on the cropping pattern before and after the advent of the canal. Section three mentions changes in sources of water for agricultural purpose and how they have impacted agriculture in the study districts. Section four presents methodology and outlines the data sources for the study. It also mentions the analysis plan and presents the hypotheses. Section five discusses the results which include trend analysis and statistical outcomes, and also discusses the implications of the findings. Section six provides concluding remarks including the limitations of the study.

The result is explained in two categories: Impact of canal water on NSA and impact of canal water on yield. The trend analysis shows an increase in NSA for wheat, oilseed and pulses in all the three districts. The means, standard deviation and growth rate in NSA for the three districts indicates that in the post-canal period all of them had positive and stable growth rate in NSA compared to pre-canal period. The trend line indicates that with canal water, yields of oilseeds and pulses have increased sharply in all the three districts, while those of wheat have not changed much. In the post-canal period, Gandhinagar had experienced higher mean yield of oilseeds and pulses than of wheat. On the other hand, Vadodara and Ahmedabad registered higher growth in yield for all three crops. Analysis of pre- and post-canal period for the selected three crops shows that in most cases yield inconsistency of crops has declined in the post-canal period owing to the timely availability of canal water.

The chapter recognises a few limitations as well. Firstly, while drawing the conclusions, it was not possible to rule out other confounders that could affect the outcome, like government policy, market environment, or incentives by government to increase production. Secondly, those factors that could have contributed to better agriculture in Gujarat, like improved market access are not captured in the present study.

Prerana Desai Vaidya in Chapter 23 discusses the results of an empirical study carried out to understand the impact of Narmada water. The study villages are from five districts—Narmada, Bharuch, Vadodara, Panchmahal and Kheda. Based on a survey of more than 1,000 households along with interviews and focus group discussions, the author concludes that access to Narmada water for irrigation has resulted in higher agricultural production and hence income. Results also show an increase in agricultural production by more than 50 per cent with three-to five-fold increases in farm income. While foodgrains and pulses cultivated ensure the households' consumption requirements, farmers have also started cultivating new varieties like Bt cotton and new crops like sunflower and seeds of various crops. Both dependability on groundwater and cost of irrigation has declined. Farmers also have access to support irrigation in case of poor rains in the monsoon. With the introduction of *rabi* and summer cultivation, the access to Narmada water has generated more employment for agricultural labourers. Wage rates for farm labourers show an increase of 40 to 50 per cent and often there is a shortage of labour. Though in varying degrees, the groundwater and drinking water situation has improved in most of the study villages and water tables have risen after the advent of Narmada water.

By increasing access to both green and dry fodder, the canal irrigation has given a boost to the animal husbandry and dairy activities. Overall, Narmada water has been instrumental in improving living condition in the study villages. However, the author also states that irrigated agriculture and more intensified activities of animal husbandry have led to a greater burden of work for the womenfolk.

Chapter 24 by Subhrangshu Goswami shows that the allocation of water for Gujarat from SSP has two distinct components—water for irrigation and water for domestic use and industry. While the irrigation component is well researched and debated by planners and policy makers at different levels, the drinking water component has been in the limelight mainly for political reasons. Once the project is completed, the drinking water component of it would cover more than 55 per cent of the villages and more than 65 per cent of the urban centres of the state. While the very need for the project has been questioned at times by some researchers and activists, an effort has been made in this chapter to re-examine the need for the Sardar Sarovar canal based water supply project in the context of the overall status of water supply in the state and the changing strategies of the state government to cope with the pressing problem of water supply. The chapter also discusses the issues and challenges involved in implementing the project and achieving its holistic objectives, in the context of its built-in-weaknesses, which arise from its concept and design on the one hand and the socio-cultural and administrative capabilities of the state on the other.

It is evident from the study that the Sardar Sarovar Canal based drinking water project in Gujarat is of immense importance. The state has been suffering from a chronic problem of drinking water supply since 1960s. Over time, with increasing demand for drinking water, the concerns have become more critical with limited availability of fresh water resources, quality problem in groundwater and inter-sector competition. With the changing nature and scale of the problem, the approach of the state government has also changed substantially over time. To start with, the state government and NGOs have experimented sufficiently in rain water harvesting (direct harvesting as well as recharge of groundwater), to meet local demand. However with climate variability, increasing demand, changing lifestyle, and sectoral competition, in many instances these efforts have proved to be unsustainable in the long run. The state government has in fact launched several quality improvement programmes for drinking water in the problem areas, the author states. Periodically the state government has prepared scarcity master plans which always focus on quick solutions. However, these solutions often proved unsustainable in long-term. Therefore, the state government, over time has started relying more and more on RWSSs. However with limited resources of fresh water and increasing demand for it over time, the existing RWSS schemes in the state are also proving inadequate.

The Sardar Sarovar Canal based drinking water supply project was launched with great hope and optimism. During the initial phase its implementation was quite slow. However, by the end of October 2010 it had covered 6,513 villages (68%), and 114 towns (87%) of the proposed coverage, which is a commendable achievement. In urban areas, the impact in terms of increased per capita supply of water and reduced dependency on groundwater is evident in many cities.

The chapter also observes that while this is perhaps the world's largest pipeline water supply project with features like financial sustainability, improved operation systems and institutional sustainability, it has several built-in weaknesses like any other long distance pipeline scheme for water transport. Arising from the project concept and design on the one hand and the socio-cultural and administrative capabilities in the state on the other, these weaknesses are of varying nature. They raise questions about the reliability and adequacy of supply, equitable inter-village and intra-village distribution, efficiency in O&M, water charges, recovery and financial viability, sustainability of local sources of water supply, and above all the capacity and efficiency of village level water management. Overcoming these weaknesses should be a major concern of policy makers and implementers at present, to achieve the long term sustainability of the project.

Chapter 25, by Dinesh Metha and Meera Metha, addresses urban drinking water security and sustainability in Gujarat. One of the key challenges of urbanisation in Gujarat, according to the authors, is provision of drinking water. A large part of the state is water stressed and has severe shortage of drinking water. The chapter is an attempt to show that with Narmada water reaching the interiors of Kachchh, North Gujarat and Saurashtra, and the Government of Gujarat's priority to this sector, there is now a possibility of drinking water security for urban areas of the state. The chapter also focuses on two major issues that affect sustainability of water—non-revenue water and cost recovery. These two factors reflect drinking water management practices and have an important bearing on overall service levels. For both these issues, the authors describe the current situation and discuss the potential ways ahead to increase sustainability of urban water supply services. In describing the existing situation, information from the studies under the Performance Assessment System (PAS) Project at CEPT University has been used.

Although the Narmada Canal has improved the coverage and increased the supply of water in ULBs, a large number of ULBs in Gujarat are still not secure in terms of their water requirement. With rapid growth of urban population, the quantum of water required from Narmada canal and other sources will have to increase manifold. The Government of Gujarat has initiated a number of schemes under its new programme for Swarna Jayanti Mukhya Mantri Shaheri Vikas Yojana (SJMMSVY) to provide water through Narmada canal wherever possible and ensure adequate sources of water supply to meet the norms of 100 lpcd of water for all ULBs. While the goals of adequacy and coverage of drinking water supply are likely to be met in Gujarat, there are issues of sustainability of these goals.

According to the authors, Non-Revenue Water (NRW) is one such issue that affects sustainability of water and is an important indicator for assessing the efficiency in service operation of a water supply system. High levels of NRW seriously affect the financial viability of water supply provision due to lost revenues; and increased operational costs adversely impact the quality of the service provided. No ULB in Gujarat has water meters, nor do most ULBs in Gujarat conduct water audit to assess the quantum of NRW. In order to gauge a better assessment of 'real' NRW, it is essential for all ULBs to undertake at least preliminary water audits. Reduction of NRW can be achieved by improved management practices of detecting leaks and repairing them quickly, identifying illegal connections and collecting revenue from them, and regular monitoring of NRW by establishing District Metering Areas (DMAs) and installing meters at supply points and at the consumer end.

Financial sustainability of a water supply system requires that ULBs cover their entire O&M costs. The cost of water supply in Gujarat is high and ULBs recover only about two-thirds of it from water tariffs and taxes. The requirement is to look at the entire gamut of operations and look for areas in which costs can be reduced and revenues enhanced. The case of Kalol municipality is described for this purpose. A preliminary water audit suggests that the NRW in Kalol is much higher than reported by the ULB. It is possible to meet the shortfall of water in Kalol by reducing the NRW. Physical losses account for a major part of NRW and the municipality would benefit immensely from plugging all the leaks and monitoring flows in the system regularly.

The chapter concludes with a note that while it is possible that Narmada canal water will be available in the foreseeable future to meet the water demand, it is necessary to take adequate measures to ensure sustainability of the urban drinking water security that Gujarat is promised. Lastly, for effective planning and maintenance all ULBs need to develop asset management systems. Only then, the authors ascertain that it will be possible to ensure that security for urban drinking water achieved at great costs is sustained for the future generations.

The Chapter 26 by R. Parthasarathy covers two distinct aspects. First a discussion on the agenda toward development of the command area as outlined in the *Planning for Prosperity* by SSNNL in 1989 is presented. Using rapid appraisal technique, two sets of field areas—one which is managed by the corporation and the other where an NGO is collaborating with SSNNL are studied.

It was evident that in both these areas there are significant changes in cropping pattern and improvements in yield and even secondary impacts like allied activities as well as raise in ground water table were evident with the advent of Narmada waters. However, in the case of the NGO assisted area there is another important achievement in terms of water management. If the basic objective of forming WUAs was to ensure equitable and efficient use of water, in the 11 NGO supported WUAs technically sound and systematic construction of sub-minors was found to be instrumental in increasing farmers' access to water in all reaches. This is in sharp contrast to the WUAs in which the sub-minors were constructed by private contractors and there was no apparent role of the WUAs in water distribution. It is evident from the various facets of implementation that the government and an NGO could form a mutually beneficial partnership. For this to build up the chapter argues that the PIM policy needs to be strengthened.

On the second aspect, the author discusses at length the other school of thought that believes in a large potential gain of involving private players in water distribution and management. He points out that a Public Private Partnership (PPP) model is also possible within an array of available management options. Indeed, the PPP model was visualised in building and managing the distribution network in the SSP command toward an idea that it would be able to take care of a substantial future expenditure on building and operating the irrigation network. In fact, the scope of PPP, as the author has pointed out would include not only financing and constructing the canals but also O&M and water distribution. The chapter discusses details of a pilot study undertaken during the year 2006 in the SSP command by CEPT and CRISIL with an objective of assessing possible changes in agricultural due to new irrigation and to ascertain what role these could have on the success of PPP and to determine the optimum structure for PPP in the pilot region.

Evidently the water management aspect has been meticulously planned for the SSP and the construction of minors and sub-minors was envisaged to be carried out by the users and the first phase of canal construction is already completed. Though there are perceptible positive changes in cropping pattern, yields and so on, the author notes that with the advent of new waters, these changes by default would be a self-directed function. Therefore, to achieve sustainable success of an infrastructure investment besides exemplary planning, SSP should also focus on water management issues considering new developments. One of them is the competing demands for water and the other is rising costs of irrigation network development. As a complementary policy to PIM, one of the possible avenues seems to be involving private players with well defined water rights. Whichever the model one pursues, the chapter argues that community involvement at the local level as a necessary condition; even as options at higher level are open to various players. In fact, there is a social limit in all economic partnerships.

In Chapter 27, Tushaar Shah and others advocate ten strong rationales for pipelining water distribution in the Narmada Irrigation System instead of constructing open channels. In support of their arguments they discuss the successful experiences of irrigation development in various districts in the Upper Krishna Basin in Maharashtra and the investments made in pipelines in many Gujarat villages. In the backdrop of farmers' resistance to provide land for constructing distribution structures below the minors the authors point out that even several years after completion of the dam and the main and branch canals, the Narmada project has been able to serve less than 10 per cent of the designed command area. It will take, the authors argue, years or even decades to realise the full benefits of the project.

Farmers, instead of providing land for surface channels, have invested in diesel pumps and flexible rubber pipes to siphon or lift water from the SSP system. This private lifting of water, to which the SSNNL has to turn a blind eye, is emerging as the dominant irrigation model. It is argued in the chapter that the formalisation of piped water distribution is possible and it can be brought within the SSP's regulatory framework. The authors consider and discuss land saving as one of the significant benefits of pipelining. Further pipelining is cost effective and water saving in terms of reducing evaporation which ultimately can contribute to the irrigation productivity of water applied. The key issue is who constructs and manages the pipeline systems: if the task is performed by government departments, chances are high that it will fail, given the dismal record of government managed pipeline schemes for irrigation dependent on gravity flow. But if the government encourages farmers and their co-operatives to construct and manage pump-based piped distribution, results can be successful as the Maharashtra example suggests. For assessing benefits and costs of pipelining, the chapter argues, it should not be compared with surface distribution systems but, given its size, SSP must be viewed against the larger canvas of the entire irrigation economy of Gujarat, which pipelining can alter in fundamentals ways.

The authors point out that a government Expert Group had summarily rejected this proposal relying on the dismal performance of government-managed piped distribution but ignoring the widespread evidence of success in private piped distribution. The authors maintain that the idea of pipelining, though criticised heavily by irrigation planners in 1970s and rejected as being unviable, has its own merits. Concluding the chapter, the authors argue that given encouragement and supportive policy environment farmers and their cooperatives will take over the entire responsibility of investing in pipeline projects and managing them. The pipelining idea has many advantages and it can throw up possibilities for institutional innovation and productivity gains. The authors perceive the chapter as a 'thought piece' written to generate a debate and facilitate more detailed analyses and calculation.

Dinesh Kumar and Nitin Bassi in Chapter 28 discuss the potential indirect benefits from SSP which were not envisaged at the time of project planning. Some of these benefits are sustainable well irrigation, reduced cost of energy for pumping groundwater in the region and improved quality of groundwater resulting from enhanced recharge. The authors claim that these benefits can outweigh the direct benefits of irrigation and hydro-power and water supply. Along with on-farm water management and equitable water distribution the SSP needs to address the challenge of maximising the economic, social and environmental benefits. They argue that in place of a conventional approach, a unique strategy is needed to realise such benefits. This will include technical and institutional interventions.

Groundwater overexploitation, and the resultant depletion and quality deterioration, have indeed become among the greatest environmental problems facing the state. They are more severe in central and north Gujarat, which are likely to get less benefit from SSP in terms of area irrigated. In such a situation, seepage from canals can be reused through wells and thus SSP can make important contributions in enhancing groundwater recharge in over-exploited areas to sustain well irrigation and to reduce well failures. Such a plan would also help in reducing carbon emissions through generation of clean energy, improvement in quality and the resultant improvement in water supply in villages and urban areas with positive health benefits. Further, they also state that improvements in the river ecosystem can be made by releasing water into the rivers of north and central Gujarat. After analysing the regional water situation they point out that SSP has the capacity to level-off regional imbalances in water availability and can change the agrarian economies of all the four regions of Gujarat. To this end, judicious allocation of SSP water among the regions is necessary. This would then translate into greater allocation of water to central and north Gujarat and Saurashtra which are the water-scarce regions. The authors support this argument by observing that that the studies in other parts of the country have confirmed that the marginal returns from irrigation are higher in water-scarce regions than in water-rich regions.

A technical approach to maximise benefits in water-scarce areas includes releasing in the monsoon, excess water through canals, which otherwise would flow into the ocean, and allowing the water to percolate down. Secondly, farmers should be encouraged to take crops that give higher returns per unit volume of water, rather than those which are water-intensive. Irrigation water supply to water rich areas of south Gujarat should be limited as groundwater balance in the region is still positive, and excessive irrigation can create problems of waterlogging and salinity. An institutional approach, as discussed by the authors, includes farmers' participation in irrigation management, volumetric pricing of irrigation water, supply rationing and tradable water rights. Enforcement of these institutional measures can lead to improvements in water use efficiency and effective demand management.

In Chapter 29 Samar Datta explores institutional alternatives to tap fishing potentials of Sardar Sarovar Reservoir and proposes a model. Given both the technical and socio-economic complexities in fisheries development over a vast water body, the NWDP Award has not given any formula for cost and benefit sharing in the context of fisheries. Therefore, in absence of a clear-cut operational policy and/or coordinated efforts towards fisheries development in the SSP, the three states concerned seem to have followed altogether different courses of action so far. While MP, holding the lion's share of this reservoir, has been pursuing fisheries independently within its physical jurisdiction with the help of local fishermen organised under the State Fisheries Federation and a contractor, Gujarat and Maharashtra do not as yet have an operational plan. Among the various institutions involved in the promotion of fisheries in the SSP, only the NCA has the mandate to resolve possible inter-state differences in fisheries development.

Before exploring institutional alternatives, in absence of any precise fish productivity figures, the author assumes an average yearly water spread area of 23,000 ha under the SSP and comes up with the net annual income figure of about Rs. 5 million in the most conservative scenario; about Rs. 12.7 million in a more liberal scenario, and about Rs. 81.7 million in the most liberal scenario. He then argues that no investor—whether private or public—is likely to invest where the returns are low, and given the poor record of cooperatives in the country, he also rules out the cooperative alternative. He maintains that it is possible to combine the advantages of a farmers' cooperative solution in the backward linkages of the value chain with the advantages of a corporate approach in the forward linkages component, namely in post-harvest management, and to gradually and in a planned manner move towards a corporate approach for realising the highest possible value out of this resource.

The chapter also identifies major stakeholders and highlights their concerns and aspirations in order to evolve an appropriate organisational format through negotiations across these stakeholders. Applying the principles of property rights and negotiating framework, the author has proposed an organisational structure named 'Inter-State Board for Management of Integrated Fisheries in the Sardar Sarovar Reservoir' (IBM-IFSSR). However, in his concluding remarks he author mentions that the proposed format in fact was presented to the party states and they found that it would confer substantial control to NCA, thereby interfering with their autonomy.

M.B. Joshi and K.D. Acharya in Chapter 30 present the uniqueness of the operational aspect of SSP canal network, which over the time has become increasingly important to help achieve the desired level of performance. The chapter starts with describing the distinct features of SSP and the canal network, functional requirements of canal operation and safety aspects. The chapter then discuses the challenges involved in operation of SSP and inter-state commitments. Lastly, the controlled volume concept (CVC) and its implications on canal structures and canal automation are presented.

The NMC, which is one of the largest lined irrigation canals in the world, is classified into Conveyance Network and Distribution Network with the former marked by many distinct features making its operation a real challenge. Safe and efficient operation of this canal network has a major role to play in deciding the performance of the project in terms of fulfilling the drinking and domestic, irrigation and industrial requirements. Unsteady flow conditions in the canal arise from operation of the control structures which may lead to untoward incidents. Certain specific drawdown criteria are thus being followed to ensure the safety of canal lining and to match variations in demand with due recognition of supply.

The SSP conveyance network, in addition to irrigation requirements, has to cater to municipal (domestic) and industrial water demand too. As a result it is subjected to frequent variations in flow requirements and 24×7 operation is imperative. Operation of the conveyance system is thus very critical and faces several challenges. Along with equitable distribution of water the system also has to meet certain inter-state commitments. The stochastic nature of both supply and demand imposes severe constraints on system operation.

The conventional method of canal operation cannot meet present day needs of timely water supply at critical stages of various crops, and often leads to a typical "tail-enders' syndrome". To meet these requirements and for successful operation of distribution system, it has been planned for flexibility in operation by controlling the volume of water in each pool. The authors describe the working of this method which virtually eliminates the need to regulate reservoirs and increases operational flexibility. SSNNL is developing in-house models to meet the specific operational requirements of SSP such as Uncontrolled Canal Siphon (without Gates), Aqueducts, in-line pumping stations and CVC.

Canal Automation has been evolved as a necessary technical solution to operate all the control structures simultaneously. It has been planned as an integral part of the planning and designing of the SSP canal network. To obtain first-hand experience it has been decided to implement it as a pilot project on about 600 km length in Phase-I of the command area, the Vadodara Branch Canal (with Jambusar and Kundhela Branch Canals) and Sakarda Branch Canal systems. After gaining this experience, the authors opine that it will be scaled up to the entire network components of more than 8.5 cumecs with suitable adaptations and modifications as may be necessary.

Connecting the Last Mile, Chapter 31 by S. Jagadeesan, sums up the gigantic project called SSP and highlights the plans of SSNNL. He states that SSP, even partially implemented, has emerged as a saviour amidst sobering scarcity of water. The Narmada canal has become a perennial water grid for Gujarat as it serves three-fourths of the state with its already constructed 26 branch canals and several distributaries. Considering the canal network down to minors, the irrigation potential developed is 503,000 ha and the project is all set to contribute—after completion of the ongoing Major and Medium projects—1.3 million hectares to the national goal of creating irrigation potential of 4.2 million hectares of land under the ambitious Bharat Nirman Yojana.

Even with partial completion of the dam and canal network, Narmada water as a harbinger of socio-economic and human development is reflected in its various benefits, tangible as well as intangible, ranging from rural water security to combating global warming. Cumulative eco-friendly generation of more than 18 billion units of electricity has put SSP prominently among the mega-hydropower projects. Results in the area under coverage indicate that Narmada water has not only provided irrigation, but also facilitated crop diversification and enhanced yield. Considering its tangible and intangible benefits such as stabilising and boosting agricultural production, creating employment in rural areas, preventing seasonal out-migration and rapid desertification, arresting groundwater depletion and improving groundwater quality, reducing infant mortality and school drop-out ratios and improving overall public health, the project authorities are obligated to expeditiously complete the remaining works so that such benefits can reach to the remaining parts as well.

In order to complete the project and the command area the author lists a set of measures adopted. An ambitious and yet achievable road map has been prepared with an overall target of completing the Project by 2014-15. This includes the remaining work of the Main Dam and works of Branch Canals, Distributor Canals, Minors, Sub-minors and structures thereon. A comprehensive implementation plan, encompassing various command area development activities, is being prepared for each area; the land acquisition process has been accelerated and joint measurement and preparation of land acquisition proposals, survey and design works has been outsourced. To ensure and monitor the quality of work in progress on a sustained basis, a third party quality consultant is being appointed. A district level committee headed by the Collector has been constituted for up-scaling PIM activities. GoG has also taken pioneering initiatives to promote Pressurised Irrigation Network System (PINS) and MIS. The Project Authorities have undertaken the exercise of exploring the potential of implementing drip irrigation practices wherever possible, considering the existing cropping pattern and its suitability for drip technology. To minimise conflicts of interest between farmers and drinking water beneficiaries, the author states that approval of the State Government has been obtained for separate feeder pipelines for drinking water supply.

Part 6

Impacts So Far

21

Irrigated Agriculture through Sardar Sarovar

N.B. Desai

Introduction

The Narmada is an inter-state river flowing through Madhya Pradesh (MP), Maharashtra and Gujarat. The terminal dam on this river has been built near village Navagam in Gujarat for creating the Sardar Sarovar Reservoir. The dam submerges lands in three states and its benefits and costs are shared by four states including Rajasthan.

Owing to differences among the basin states the matter was referred in 1969 to the Narmada Water Disputes Tribunal (NWDT) for adjudication and it gave its decision in 1979. The project was formulated in accordance with the decision of the Tribunal and presented to the Government of India (GoI) in 1980. It received clearance from the environmental and forest angles in late 1987 and investment clearance of the Planning Commission in October 1988. The works of the dam, two power houses and canal system were started from 1987-88.

The main decisions of the tribunal were:

Allocation of water at 75 per cent dependability

Madhya Pradesh	:	18.25 MAF (million acre feet)
Maharashtra	:	0.25 MAF
Gujarat	:	9.00 MAF
Rajasthan	:	0.50 MAF
		28.00 MAF

Height of the dam	:	FRL 455 ft
FSL of the canal	:	RL 300 ft
Allocation of Power		
Madhya Pradesh	:	57%
Maharashtra	:	27%
Gujarat	:	16%

The project was estimated to cost Rs. 64.06 billion at 1986-87 price levels. The important features of the project are:

- A concrete gravity dam, with FRL 138.68 m (455 ft) and MWL 140.21 m (460 ft) to be constructed across Narmada near Navagam village in Gujarat.
- Underground River Bed Power House with installation of 6 reversible turbine generators units of 200 MW capacity each.
- Canal Head Power House with installation of 5 Kaplan Turbine Generator units of 50 MW capacity each.
- 458 km long Narmada Main Canal (NMC) with a discharge capacity of 1,133 cumecs at its head, tapering down to 71 cumecs at the Gujarat Rajasthan Border.
- 75,000 km distribution canal network for an annual irrigation of 1,792 million hectares.

Canal System

The entire irrigation network of the canal system is divided into two groups—the conveyance network and the delivery network. The conveyance network consists of main canal, branch canals and distributaries. The delivery network consists of minors, sub-minors and field channels.

The NMC, one of the largest lined irrigation canals in the world takes off from Pond No. 4 downstream of the canal head power house on the right bank of the Narmada. It is aligned as a contour canal with a length of 458 km up to Gujarat-Rajasthan border and a further length of 74 km in Rajasthan.

There are 38 branch canals off-taking from the main canal at various locations along its 458 km length with discharge capacity ranging from about 2 cumecs to as high as 425 cumecs. The major branch canals are Miyagam, Vadodara, Dholka, Saurashtra and Kachchh, with discharge capacities at head of 144, 76, 57, 424 and 220 cumecs respectively.

The minors in village service area carry water to *chaks* of size varying from 25 to 65 ha, sub-minors off-take from minors and carry water to sub-*chaks* of 8 ha. Field channels off-take from sub-minors and deliver water to individual farms.

Delivery Network

For planning the delivery network the entire command area is sub-divided into small basic units of 200 to 500 ha known as Village Service Area (VSA). One revenue village would generally form one VSA. Each VSA would be further sub-divided into smaller units of *chak* (25-65 ha) and sub-*chak* (6-10 ha). The entire canal system from reservoir to sub-*chak* level of 8 ha will be lined.

Within a service area an unregulated free flowing system would carry water to all *chaks* (25-65 ha) during the weekly allocation period. The delivery within a *chak* would be through a regulated free flow system, the *chak* supplies being rotated at full *chak* flow to the farmers. The farmers will be free to select crops, cropping pattern and method of irrigation for utilisation of the water allocated to the service area except that the system would not be operated to serve the special needs of perennial crops.

Canal Automation

The project envisages extensive rather than intensive irrigation to cover as much of the drought affected area as possible in spite of the limited quantum of water available. The farthest field will be about 700 km from the headworks. Canal automation is proposed to provide quick, reliable and equitable response to water demands of even tail-enders and to avoid wastage of precious water.

In controlled volume operation adopted for the canal system the volume of water contained in each pool of water between two adjacent cross regulators can be altered as desired to accommodate operational needs. Though the flow in the canal will vary throughout the year in response to scheduled deliveries, the volume of water within each pool will remain almost equal to that at full discharge. To meet minor mismatching and short term changes in local deliveries, water will be drawn out of or placed into storage within the respective pools as conditions dictate.

A Feasibility Report on control and communication system for canal automation has been prepared by Gujarat Communication and Electronics Ltd. The canal system under construction has been designed on a controlled volume concept. All control structures are also designed for canal automation. The automation is proposed to be introduced in phased manner so that problems, if any encountered initially can be taken care of in the subsequent stages.

Command Area and Crop Pattern

The command of SSP covers a very large area with varying agro-climatic conditions. It has, therefore been divided into 13 regions considering the ultimate objective of choosing feasible sets of crops for each region, regional allocation of water and its efficient management, and regional potentials and constraints in conjunctive use of canal and ground water.

A very important feature of the climate of the command is actual precipitation, both in total quantity and distribution, which varies greatly from year to year, and crops suffer largely as a result of moisture stress. It is estimated that almost three-fourths of the command is drought prone.

For planning of crops and water use in the command area it is important to know broad soil types and land irrigability classification of all its regions. On the basis of available data on classification of soils and land irrigability classes, it is found that all regions are good for irrigation except some critical areas in two regions—Region 4 in Vagra and Jambusar talukas of Bharuch District and Region 7 in Dholka and Dhanduka talukas of Ahmedabad District close to the Gulf of Khambhat. These regions require special plans for water use.

Cotton, bajra, jowar, wheat and paddy are major crops in the command. Cotton is the predominant crop. In irrigated areas, it has continued its dominance, but bajra and jowar, the low value cereals, loose their places in favour of high value cereals paddy and wheat.

The creation of irrigation facilities has motivated farmers to adopt technological improvements and innovations to increase agricultural production. The availability of short duration varieties of paddy resulted in replacement of long duration unirrigated varieties of paddy, and the early release of land under paddy, coupled with irrigation availability and high yielding varieties of wheat, have encouraged farmers to expand areas under irrigated paddy and irrigated wheat.

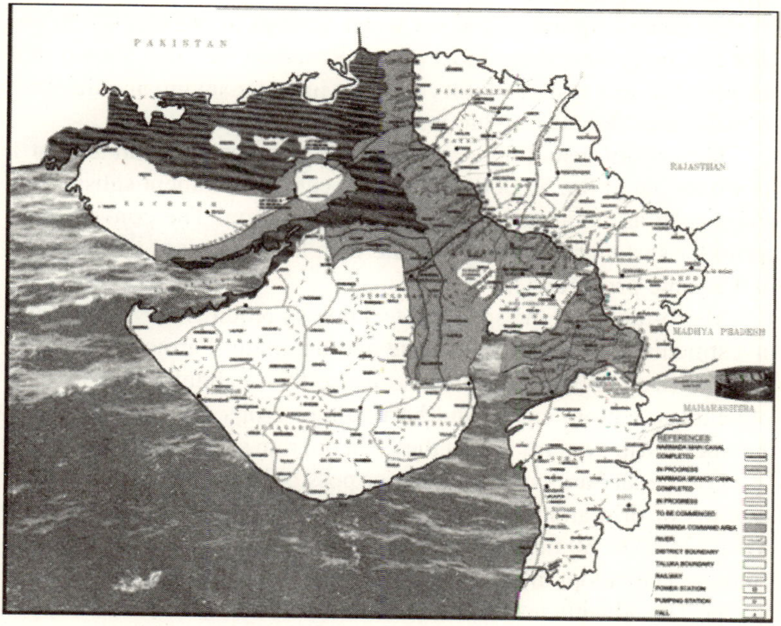

Map: Alignment of Narmada Main Canal and Command Area of SSP

Source: SSNNL, Gandhinagar.

A notable feature of the agricultural scene in Gujarat and in the Narmada command is that even the marginal and small farmers allocate 25 to 40 per cent of the gross cropped area of their holdings to commercial crops. It may also be pointed out that the cropping intensities achieved by marginal and small farmers in most of the regions are higher than those of other cultivators.

Micro-Irrigation Systems

The physical efficiency of water use in traditional irrigation practices such as small border irrigation is low in the well-irrigated commands of regions such as Saurashtra, Kachchh and North Gujarat due to heavy evaporation and percolation losses. Irrigation water being very scarce in these regions, the physical efficiency of water use and thereby its economic efficiency need to be improved to manage water demand for agriculture.

By using drip irrigation, net water savings as high as 40-50 per cent can be achieved for most crops. Research has also shown that adoption of drip system results not only in water saving and increase in yield rates, but also reduction in input costs.

The cost of installing a conventional drip system in an area of 1 ha works out at Rs. 50,000 with 50 per cent government subsidy for crops such as sugar cane, vegetables and cotton. The conventional sprinkler systems, which also come with a government subsidy, fall in the same price range.

Micro-tube drip and sprinkler technology requires very low capital investments and is easy to operate and maintain. Farmers with very small as well as large holdings can use this technology to irrigate vegetables, cotton, castor and horticulture crops. The advantage of the micro-drip systems is that they can operate at very low pressures.

For Sardar Sarovar there are three possibilities for micro-irrigation systems of drip and sprinkler: (1) at 300 to 400 ha Village Service Area, (2) 40 to 60 ha *chak*, and (3) 6 to 8 ha sub-*chak*. The present thinking is to adopt micro-irrigation systems at sub-*chak* level that is 6-8 ha. The advantages of this are: (i) easy O&M systems and (ii) the possibility of having small group of contiguous farmers adopting a homogeneous cropping pattern.

SSP may consider providing subsidies and incentives for introduction of micro-irrigation systems at sub-*chak* level. When micro-irrigation systems are adopted on a large scale suitable Public-Private Partnership may be considered for adoption.

In 2007 SSP has adopted a policy, for a period up to 2012-13, of water rates for irrigation in its command area. The policy provides that the rates to be charged for drip and sprinkler systems would be 35 and 65 per cent respectively of normal rates per hectare of relevant crops.

Canal Automation-Pilot Project

For efficient but flexible safe operation with reduced response time, the entire canal network down to 8.5 cumecs involving about 2,500 km of canals and about 1,400 control structures is planned to be operated by a Supervisory Control and Data Acquisition (SCADA) based Remote Monitoring and Control System.

Command area development works and gate works of branch canals, distributaries, minors and sub-minors are planned to be completed expeditiously. Therefore, it has been decided to implement a pilot project for Canal Automation at Vadodara Divisional Operation Centre which will include Narmada Main Canal Reach 0.0 to 105.344 km, Vadodara Branch Canal with Jambusar and Kundhela Branch Canals, and Sakarda Branch Canal.

Telecommunication Consultants India Limited (TCIL) has been awarded a consultancy assignment for Reviewing Detail Design and Bid Documents for Canal Automation Pilot Project and preparing the final version of bid document so that implementation of the pilot project can be taken up. Tenders for implementation of the Canal Automation Pilot Project are planned to be invited in the first quarter of 2011.

Conjunctive Use

Appreciating the importance of conjunctive use management towards agriculture and environmental health of the command area SSP has integrated specific strategies. As the first step, credible estimates of region-wise groundwater to be used conjunctively and that to be disposed of from partially bad and bad areas, were made by the project. These estimates give 2.71 MAF of useable ground water and 0.66 MAF of unusable groundwater to be wasted by pumping.

The farmers and farmer organisations are expected to develop new and additional groundwater in the VSAs through tubewells, borewells and open wells because of limited surface water allocations. The present groundwater use in the command area is 980 MCM. The farmers would be fully supported in their efforts and bankable projects may be prepared and some incentives considered by the project.

Where groundwater development at VSA level is not viable the agency in charge of management will drill and operate the tubewells and open wells as required. Such areas could be hard rock areas of poor yields where farmers may not be able to develop tubewells and open wells. Moreover, there are saline areas where groundwater has to be wasted and pumped out.

In general, as soon as water table is about to rise to 5 m below ground level in the command, the Agency will step in to ensure that appropriate arrangements for conjunctive use or pumping for disposal outside are set in motion so that significant waterlogging is not allowed to develop in the command area.

The experience of the Mahi Project demonstrates that farmers do come forward for conjunctive use once they have realised the value of water in their crop production systems. For realising this objective education and training of farmers has been clearly recognised by the project. Gujarat has a well developed Water and Land Management Institute at Anand along with three regional centres which are equipped for providing training to farmers besides irrigation and agricultural engineers.

Drainage Aspects

SSP is planned and being implemented with an integrated framework of vertical as well as horizontal drainage each fulfilling its technical and environmental role. Vertical drainage is through pumping for conjunctive use and disposing of saline water.

The project has recognised that the requirements of drainage will vary from region to region and also within a region as adequate drainage in a particular area/region depends on a number of factors like topography, storm rainfall pattern and intensity, types of soils, and types of crop grown. Against this background SSP in 1994 got experts to prepare a Manual of Surface Drainage Systems. The manual provides guidelines for planning and design of drainage systems.

The project has also proposed use of Geographical Information System (GIS) and Remote Sensing Techniques (RST) to study *inter alia* drainage aspects of the command area. The Bhaskaracharya Institute for Space Applications and Geo-Informatics under the Department of Science and Technology, Government of Gujarat has furnished a Geographical Information System for SSP Command Area

Phase I Project proposal (2009) to SSNNL. It suggested the creation of a digital database of various cross-sections of drainage structures for Phase I Narmada Main Canal 0-144 km and such a database would be integrated with catchments and micro-watersheds. Severely waterlogged and inundated areas during the last five years' heavy downpour events would be derived from satellite imagery and integrated with the digital database of Phase I command area. To start with a pilot study for Command Area Phase I representing an irrigation block of 5-8 thousand hectares is to be taken up by the project. The work components and estimates of the whole project would be evolved on completion of the pilot study.

For tackling the problem of rise in water table and subsequent waterlogging, a drainage study for the command area of 446,000 ha in Phase I has been done in 1981. For preparing a comprehensive drainage plan of Phase I consultants were engaged to carry out field surveys and micro-level planning and to prepare block-wise drainage plans in 1990-91. These have now been reviewed and modified taking into account developments in the command area. Using such drainage plans, surface drainage works would be taken up in selected areas depending upon rise in water table resulting in waterlogging.

Project Status and Benefits

The work of raising the concrete dam up to the crest level of spillway EL 121.92 m has been achieved. About 6.577 MCM of concrete has been placed in the dam against a total estimated quantity of 6.820 MCM representing 96.43 per cent physical progress. The work of constructing piers, installation of radial gates and bridge remains to be completed. This will be done in three years after permission is received from the NCA, upon completion of what is left of the R&R.

The Canal head power house comprising five conventional Kaplan turbine generator units of 50 MW each was commissioned between August 2004 and December 2004. Up to October 2010 it has generated about 1,920 million units of energy. The riverbed power house comprising 6 units of reversible Francis turbine generator units of 200 MW each was commissioned between February 2005 and June 2006. Up to October 2010 it has generated about 15,511 million units of energy.

The canal system comprises 458 km main canal, 2,585 km branch canals, 5,112 km distributaries, 18,413 km minors and 45,058 km sub-minors besides field channels to develop irrigation potential of 1.792 million ha. The progress on these canal components has been 100, 71, 30, 27 and 21 per cent respectively. The potential of 3.36 lakh ha along with canal system up to sub-minors has been developed so far.

The project is also under the Accelerated Irrigation Benefit Programme (AIBP) of the Ministry of Water Resources, Government of India which provides funds for construction of the canal system. The project commands a vast area and the farmers in the command area, which is drought prone, to a large extent use waters through sub-minors wherever available, and also by installing pump units from minors, distributaries and branch canals and also from some locations on the main canal. It is, therefore, difficult to assess actual utilisation of water for irrigation in the command area.

The Ministry of Water Resources, Government of India has got done an assessment of Irrigation Potential created in AIBP Funded Irrigation Projects in India, using Cartosat Satellite Data 2009 through National Remote Sensing Centre, Hyderabad. Identification and mapping of existing irrigation network with Main Canal, Branch Canals, Distributaries, Laterals and Sub-Laterals of SSP has been done from satellite image on computer screen using ERDAS Imagine image processing—Arc GIS software through on screen digitisation by displaying the image at 1:4000 to 1:2000 scales. In addition to the canal network, the associated irrigation and drainage structures have also been identified and inventoried. The digital map base for the study area of Phase I command with rivers/streams, roads/rails and villages using high resolution data has been prepared and used further for preparing map outputs.

Inventory of information in terms of number of canals comprising main canal branch canals, distributaries, minors, sub-minors, their chainages and respective lengths have been computed for the AIBP component area of SSP. The status of canals was also assessed in terms of 'completed' or 'uncompleted' or 'pending', based on satellite data inventory information. This inventory has been compared with information obtained from the field. Based on these data the irrigation potential was assessed along with balance irrigation potential to be created, as of April/May 2007.

All the canal lengths and their off take chainages from parent canals derived from satellite data are found comparable with reported field statistics, namely canals which are fully completed, under progress, and not taken up so far, except two minors.

The irrigation potential is assessed using the irrigation infrastructure derived from the Cartosat Satellite data. Based on the assessment of physical progress, the overall irrigation potential created in Phase I of the project is assessed as 392,158 ha as against the Ultimate Irrigation Potential (UIP) of 446,609 ha. The balance irrigation potential to be created is estimated to be 54,451 ha as of April/May 2007.

The satellite imagery based created irrigation potential of 392,158 ha under SSP Phase I was more than the field reported irrigation potential as of March 2007 of 340,724 ha.

The Bhaskaracharya Institute for Space Applications and Geo-Informatics, Department of Science and Technology, Government of Gujarat has also studied IRS Satellite Imagery of December 2008/ January 2009 and assessed that area irrigated is 374,606 ha against reported irrigation potential of 255,000 ha in Phase I of SSP. The field engineers have found that about 60,000 out of 374,606 ha of irrigated area is of other projects including tubewells and open wells. After deducting this area of 60,000 ha, the area irrigated by Narmada Waters is 314,000 ha. The area reported by the field officers is only 96,325 ha. The BISAG studies are for mid-Rabi season and irrigated area assessed by them is fairly representative.

For Phase II command area of 1,399,000 ha NRSC studies have shown that 267,430 ha of area was irrigated in September 2009. All this area may not be irrigated by Sardar Sarovar. It may include rain-fed Kharif crops and also some area irrigated by other projects.

Canal Water and Agriculture

Prerana Desai Vaidya (2010) has given a field evaluation of the impact of SSP water based on a study carried out in 2006 and 2007. Funded WWF-India, it selected 25 villages from four of thirteen regions, namely 1, 2, 4 and 5 of the command area. These selected villages represent hard rock and predominantly tribal area and medium black and brown soil suitable for ground water development, higher agricultural development with cotton growing area, fertile land but saline ground water with acute shortage of drinking water, and sandy soil and high irrigation intensity. The focus was on study of agricultural profile of selected villages before and after SSP Canal Water.

Cultivation in all regions irrespective of the size and landholding has started with availability of canal water. Before the canal, tobacco was the preferred crop but now canal irrigation has made winter crops (Rabi) possible in all surveyed villages in addition to the Kharif crop. Wheat is now the prominent crop in all the four regions studied. With canal irrigation, cultivators grow Rabi maize. Favourable conditions have been created for optimum cropping pattern and higher cropping intensity. This has led to a rise and stability in incomes.

The change in level of income with development of irrigation has been very positive. About 80 per cent farmers, even in tribal areas, now avail of canal irrigation, taking more than one crop in a year. The gross production and yields of crops have increased by 50 to 100 per cent.

Small and marginal farmers have also started availing of full benefits of irrigation water. The big farmers have started experimenting with new crops like cotton, Bt cotton, sunflower and even started farming for foundation seeds for various crops. They provide seeds to the nearby companies situated in the region.

After the advent of canal water, rates of daily wages have increased by 40 to 50 per cent. Specially trained adivasi groups of labourers take a foundation crop of cotton, which is a highly skilled job. They are in great demand with increase in area under cotton production and increase in yields. They can now earn Rs. 300 to 400 per day during the picking of cotton buds.

According to Vaidya (2010) the priority of farmers, with increased income due to of canal water, is renovation of their houses. The second priority is to buy milch animals and bullocks. They now wish to join milk cooperatives. Next priorities are procuring equipments like tractors, threshers, diesel pump sets and motor cycles. They now desire to have electricity connections.

Issues of Debate

One of the accusations levelled against SSP is overestimation of water availability. The availability of water in the Narmada at Sardar Sarovar was evaluated using long term data from 1891 and agreed to by all partner states. CWC reassessed the amount using more than 100 years data from 1891 to 1992. M. Dinesh Kumar (2010: 174) observes that analysis of historical data of point rainfall in locations across the Narmada Basin shows that there has not been any statistically significant trend in rainfall in the region.

Another issue raised is that SSP would benefit rich farmers while the tribal population being displaced would suffer. In the villages of Tilakwada and Naswadi talukas with predominantly tribal population, where land holdings are marginal, SSP water has been made available from 2002 onwards. Desai Vaidya (2010), based on her fieldwork, that about 80 per cent farmers are fully availing of canal water facility.

It is also alleged that sugar factories have been sanctioned in the command area and canal water will not reach distant areas of North Gujarat, Saurashatra and Kachchh. Desai Vaidya's work, done in 25 villages in four of thirteen regions, shows that no sugarcane is being grown there. The predominant irrigated crops are paddy, cotton, tuver, wheat and bajra.

Another unfounded apprehension is that the irrigation efficiency assumed is unrealistic and actual coverage would be lower. Sardar Sarovar has planned lining of the canal system to 8 ha sub-*chak* level, canal automation and volumetric distribution of water, and supply of water to groups of farmers and not individuals. Projects in Gujarat have demonstrated efficiency in water management under conventional systems by capturing national productivity awards. We have to change our mindset and believe in ourselves that India can achieve high efficiency.

Some critics have argued that assessment of groundwater and conjunctive use are not realistic. The assessment was done after detailed groundwater surveys and mathematical modelling for estimating the recharge in the Phase I command area through the specialised international agency Mott MacDonald. Even in the Mahi command canal irrigation is 150,000 ha while groundwater irrigation is about 175,000 ha. The conjunctive use considered in Sardar Sarovar is quite realistic.

The most critical issue raised against Sardar Sarovar is that it is economically unviable. Irrigation projects with a benefit cost ratio higher than 1.5 are considered economically acceptable. SSP, worth an estimated Rs. 64.06 billion at 1986-87 levels and benefit cost ratio higher than 2 was given investment clearance by the Planning Commission in 1988. Economic appraisal by the Tata Economic Consultancy Services, Mumbai in 1983 indicated that the project would confer substantially larger benefits than the cost involved, with a benefit cost ratio of 1.84 and internal rate of return of 18.3 per cent. The Staff Appraisal Report of the World Bank also estimated the economic rate of return at 13 per cent in 1985. The World Bank again undertook economic appraisal based on 1991-92 prices. The results of this exercise reported in the Project Completion Report prepared in 1995 show an economic rate of return of 12.2 per cent.

The Revised Cost Estimates of SSP for Rs. 392.405 billion at 2008-09 prices were submitted to CWC by the Government of Gujarat. CWC calculated a benefit-cost ratio of 1.63 excluding revenue receivable from domestic/industrial water supply and fisheries. The Planning Commission has given further investment clearance in May 2010.

22

Impacts of Narmada Water on Agriculture in Gujarat

Amit Garg, Swati Pathak, Neha Jhala and *Bhushan Kankal*

Introduction

Gujarat state houses about 5 per cent of India's total population and 60 per cent of Gujarat's worker population depends on agriculture. But still the growth of the agriculture sector in Gujarat during 1970-1980 was slow compared to the manufacturing sector. The share of agriculture in the state's domestic product was 42 per cent in 1970-71 which fell to 32 per cent during 1980-81.

Almost 50 per cent of the area of Gujarat is cultivated. The fluctuations in net sown area (NSA) occur mainly on account of drought when sowing of crops cannot be done in the arid areas in the northern part of the state and lands have to be kept fallow. The land resources of the states have been exhausted and further addition to them is not possible without reclamation of waste lands. Annual rainfall ranges from less than 300 mm in the North West to over 2000 mm in a small area in the south. Thus a large part of the state receives low rainfall, which characterised by large variations over time and space.

The Sardar Sarovar canal system was designed to enhance the irrigation coverage in Gujarat and parts of Rajasthan and Maharashtra. The Sardar Sarovar Project (SSP) will irrigate 1.845 million ha of land, covering 3,112 villages of 73 talukas in 15 districts of Gujarat.

It will also irrigate 75,000 ha of land in the strategic desert districts of Barmer and Jalore in Rajasthan and 37,500 ha in the tribal hilly tract of Maharashtra through lift irrigation. It was also projected that new area would come under cropping, and crop-mix would change along with higher yield per crop due to irrigation water from Sardar Sarovar canal system. In the longer term it would also have a positive impact on ground water tables in the vicinity, leading to more sustainable living in this part of India. The Sardar Sarovar Dam, from which the canal system starts, is on the Narmada River in Gujarat state, 170 km upstream from where the river flows into the Gulf of Khambhat in the Arabian Sea.

This chapter examines whether Narmada water has brought newer areas under agriculture and enhanced crop yields. The overall objective of this study is to gain a deeper understanding of the impact of Narmada water on agriculture in three districts of Gujarat—Ahmedabad, Gandhinagar and Vadodara—three of the eight districts have started receiving the Narmada water from 2005 onwards (SSNNL, 2010). This is done through (1) analysing the impact of canal water on yield of three major crops and (2) analysing the impact of canal water on NSA. The chapter is organised in six sections. After a brief background of Sardar Sarovar Canal in Section One, Section Two presents the cropping pattern before and after the canal became functional for these districts. Section Three mentions changes in sources of water for agriculture and how they have made an impact on the study districts. Section Four presents the methodology part of the chapter and also outlines the data sources for the study. It also mentions the analysis plan for this study and presents the hypotheses. Section Five discusses the results, which include trend analysis and statistical outcomes, and also discusses the implications of the findings. Section Six provides concluding remarks with the limitations of the study.

Section 1 – Command Area

The gross command area of the project is 3.43 million ha, of which the cultivable command is estimated to be 2.12 million ha. The command lies between 21° 41′ to 24° 30′ N latitude and 68° 20′ to 73° 45′ E longitude, encompassing fifteen districts—Narmada, Bharuch, Vadodara, Anand, Panchmahal, Kheda, Ahmedabad,

Gandhinagar, Mahesana, Patan, Bhavnagar, Surendranagar, Rajkot, Banaskantha and Kachchh. The total number of the talukas of these districts, wholly or partially covered in the command is 73 and about 3112 villages of these talukas are expected to be served by the project for irrigation. The command is broken up into 13 homogenous agro-climatic regions taking into consideration annual rainfall, land topography, soils, ground water quality and depth and drainage conditions.

Out of these 13 regions, five are so far benefited by Narmada Canal water. Almost eight districts are included in these five regions: Anand, Ahmedabad, Vadodara, Kheda, Gandhinagar, Narmada, Panchmahal and Bharuch. Out of these eight districts, our study has been done on three important districts receiving canal water, which have been using larger share of the Narmada canal water for agricultural irrigation purpose. Figure 1 represents the boundaries of the 13-agro climatic regions in Sardar Sarovar command area, the boundaries of the all districts included in five regions getting canal water and districts of our interest for this chapter.

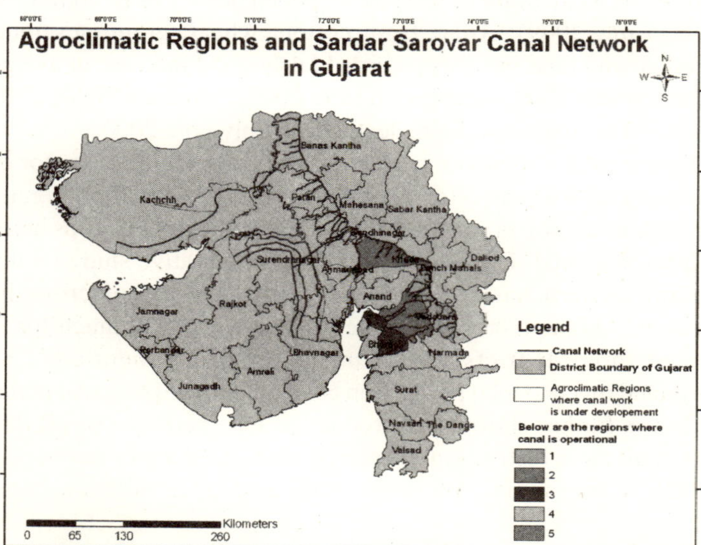

Figure 1: Sardar Sarovar Command Area and 13 Agro Climatic Region

About 75 per cent of the command area in Gujarat is drought prone while the entire command (75,000 ha.) in Rajasthan is drought prone. Assured water supply will make this area drought proof (SSNNL, 2009).

Currently land irrigated as a percentage of total cropped area (irrigation intensity) in Gujarat is considerably low, with wide variations. For instance it is 5 per cent in the Vagra-Jambusar region, 36 per cent in the Sinor-Vadodara region and averages 16 per cent in the command area. This is projected to increase up to 68 per cent on an average after the implementation of SSP (SSNNL, 1988). This will play a major role in reducing fluctuations in Gujarat agriculture.

Section 2 – Cropping Pattern Before and after Sardar Sarovar Canal

Cotton, wheat, paddy, oilseeds, pulses, jowar and bajra are major crops in the command area. Cotton has recorded the most increase in average production during 2000-2008. The cotton boom in Gujarat has been aided by the availability of reasonably priced quality of Bt cotton seed (Shah *et al.*, 2009). Apart from cotton, wheat, gram, sesamum, groundnut and total oilseeds recorded fairly good increase in average production after 2000, compared to 1999-2000 (Pathak and Shah, 2010). Production of maize and bajra declined sharply post-2000.

Based on this scenario, we have taken three major crops for our study purpose—wheat, oilseeds and pulses. The bar chart in Figure 2 shows the total crop production for the abovementioned crops before and after the canal. It actually triggers a comparative study on how cropping pattern has changed and how the share of crops has increased. Although cropping pattern has not changed much for the selected three districts, the share of the crops mentioned has contributed much to total production compared to the pre-canal period. The figure further shows that production has increased of all three crops in all three districts in the period after canal water has become available to the farmers. We discuss the causality in further sections.

Section 3 – Sources of Water

Agricultural practices in the command area depend on water availability. There are four water-related factors which might play a

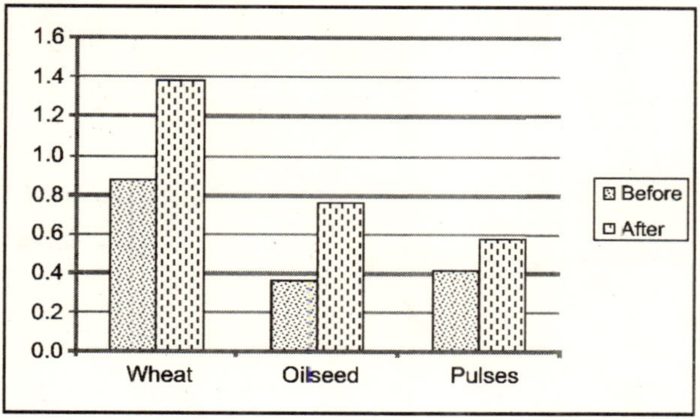

Figure 2: Cropping Pattern in Selected Three Districts

crucial role in agricultural development in our study districts—natural rainfall, surface water from the Narmada, surface water from other canals and recharge in groundwater. Amongst these four factors, our interest is to see the effect of Narmada canal water. However, we cannot ignore the effect of the other three factors during the study period.

One of the primary ways in which humans use water is by planting crops in places where they can capture natural rainfall for rain-fed agriculture. Figure 3 shows that rainfall has increased in the study area during pre-to post-canal period (CMIE, 2010). Ahmedabad and Vadodara experienced respectively (61% and 63%) average growth rate of annual rainfall from 2001-2008. Gandhinagar recorded an average rainfall increase of almost 130 per cent during pre-to post-canal period (CMIE, 2010). However, this increased mainly during monsoon period used for khariff crops which were getting sufficient water anyway. Therefore increase in rainfall would not directly affect impacts due Narmada canal water during the entire cropping period.

Changes in rate of electricity consumption in the agricultural sector during pre-to post-canal period has reduced. Although the pre-canal period witnessed the highest electricity consumption share in the agri-sector in Gujarat, in the post-canal period it falls sharply (Modi, 2011). This indicates that usage of ground water for irrigation may have decreased during post-canal period.

On the third aspect of alternative water source, there been no substantial changes in command coverage by other canal systems

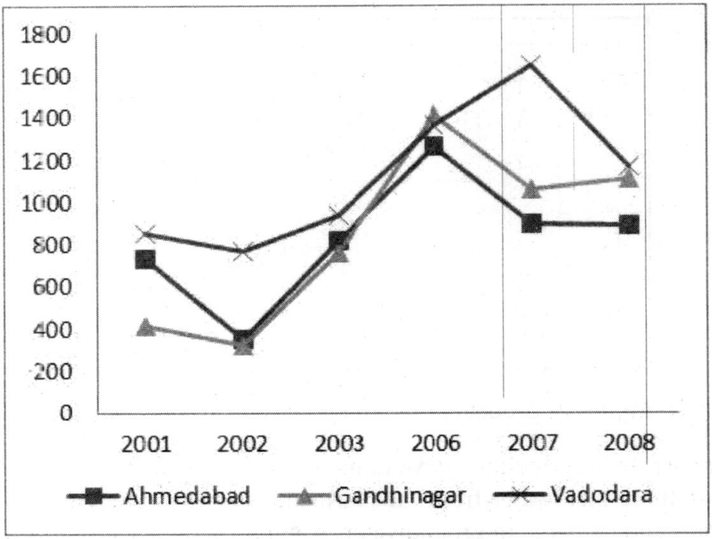

Figure 3: Trend in Total Rainfall (mm, annual)

during the study period in any of the studied districts. Narmada canal irrigation was the only change in water availability. Therefore, the effect due to changes in other systems would be negligible as compared to the effect due to changes in Narmada Canal irrigation system. The impacts of the Canal have been estimated based on a detailed methodology, considering two hypotheses, explained in the following sections.

Section 4 – Methodology

A study, quantifying impact of Narmada water on agriculture using time series data, was conducted on the three districts, selected as they were mainly fed by Narmada canal water and also included in major five agro-climatic regions. Further, using GIS, the percentages of the area of each of the three selected districts covered by the five regions were obtained. It shows that 60 per cent of total geographical area in Vadodara district is covered in regions 1 and 2 and less than 1 per cent is under regions 3 and 5. In Gandhinagar 10 per cent of total geographical area and in Ahmedabad 14 per cent is under region 5.

We have taken a 10-year time span for this analysis, dividing it in two periods—pre-canal (2000-2005) and post-canal (2006-2010). The study compares these two periods to analyse the beneficiary impact of canal water in agricultural practices in the three districts.

The cropping pattern in these three districts is also a criterion to select the number of crops for our analysis. Considering the cropping practices and their share in total production during the study period, we have taken three crops as the main crops for these districts—wheat, oilseeds and pulses.

Analysis Plan

The entire analysis was done in two parts. We did trend analysis to see how agricultural practice in Sardar Sarovar command area has responded to the availability of Narmada water. We have identified two factors to be analysed for this particular study period (2000-2010) and trend charts for each of them will give us a deeper insights about how the yield and NSA change during this time period. The average value of each of these factors for pre-and post-canal period were calculated to quatify the changes, and standard deviation for them helps us figure out the stability in agriculture due to canal water. To validate the result from the trend analysis and our two hypotheses stated below, we have done an independent sample *t* test on the mean of total yield and total NSA for all three districts. For this analysis we have taken a time dummy variable assuming 0 for pre-canal period and 1 for post-canal period. To conduct the statistical *t* test we have used SPSS statistical software.

Hypotheses

Irregular and uneven rainfall in Gujarat was a constraint for farmers to earn their living. The direct consequence of this was the use of groundwater lifted through private or public tubewells. These consume diesel, imposing the burden of a huge cost of cultivation on farmers. In this case, spread of irrigation through Narmada canal water is expected to be a major driver of agricultural growth in terms of high yield, increase in NSA, and crop production. It is also assumed to have a strong impact on cropping pattern in this particular region and in a broader way for almost all the regions fed by Narmada canal water. Given that the soil quality and ground water level have not changed much during our study period, we expect assured availability of canal water could give a very clear picture of cropping practices in Gujarat. Through this study we want to test whether, timely availability of Narmada water in agriculture causes increase in crop yield, NSA, and hence has overall positive impact on agriculture. We have tested two hypotheses as mentioned below:

(1) Assured water availability from Narmada canal causes significant increase in NSA in post-canal period.
(2) Assured water availability from Narmada canal causes significant higher yield for almost all crops in post-canal period.

The data are obtained from Centre of Monitoring Indian Economy (CMIE), Krishi Bhavan, Gandhinagar and Directorate of Economics and Statistics (DACNET). They contain information on yield (kg/ha), NSA (in '000 ha) and total production (in '000 tonnes). It must also be mentioned that based on data availability for wheat, we have used net irrigated area from Krishi Bhavan. These data also provide information on the canal network of the Sardar Sarovar command area, district boundary, command area boundary and 13 agro-climatic region boundaries.

Section 5 – Results and Discussions

The result section is explained in two categories: Impact of canal water on NSA, and yield.

Impact of Canal Water on Net Sown Area of the Crops

Increase in NSA plays a key role in increasing crop production. The most important factor to enhance NSA is assured water availability. Figure 4 shows the trend analysis of NSA for all three crops for pre-and post-canal period. The trend analysis follows an increase in trend in NSA for all the three crops in all three districts. The highest growth rate is observed for oilseeds in Gandhinagar

(168%) followed by Vadodara (64%). Gandhinagar (86%) and Vadodara (74%) experienced highest growth rate in NSA for wheat compared to Ahmedabad (47%). For pulses Ahmedabad (45%) and Gandhinagar (39%) experienced maximum growth rate.

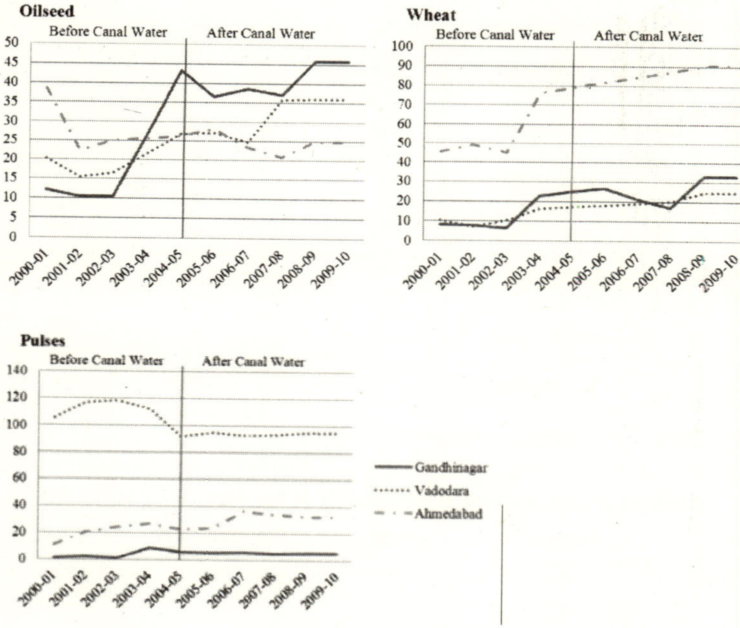

Figure 4: Trend in Net Area Sown (000ha) in Pre and Post Canal Period

The above analysis is summarised in Table 1, which represents mean, standard deviation and growth rate for NSA for all three districts. Standard deviation in the following table indicates that in the post-canal period all the three districts had positive and stable growth rate in NSA compared to pre canal period.

Table 2 represents the independent sample *t* test results for mean total NSA in all three districts. Result shows that except pulses all other crops have got statistically significant results, indicating that there has been a significant increase in NSA due to availability of Sardar Sarovar canal water. All three crops also experienced higher increase in NSA.

Table 1: Mean, SD, and Growth Rate in NSA of Selected Crops in Three Districts

District		*Gandhinagar*	*Vadodara*	*Ahmedabad*
Wheat (irrigated)	Mean–Pre		12	59
	(SD)	(9)	(4)	(17)
	Mean–Post	26	21	87
	(SD)	(7)	(3)	(4)
	Growth Rate (%)	86	74	47
Oilseeds	Mean–Pre	15	18	27
	(SD)	(7)	(3)	(13)
	Mean–Post	40	30	21
	(SD)	(4)	(5)	(0)
	Growth Rate (%)	168	64	–21
Pulse	Mean–Pre	4	109	21
	(SD)	(3)	(11)	(6)
	Mean–Post	6	94	30
	(SD)	(0.4)	(1)	(6)
	Growth Rate (%)	39	–13	45

Note: Mean NSA and SD in thousand hectares.

Table 2: Independent Sample t Test for Net Area (000 ha)

	Dummy	*N*	*Mean*	*Std. Deviation*	*t-value*
Wheat	0	5	85	30	0.01**
	1	5	134	13	
Oilseed	0	5	68	19	0.02**
	1	5	97	9	
Pulses	0	5	134	8	0.44*
	1	5	130	6	

* Significant at 95 per cent Confidence Interval.
** Significant at 99 per cent confidence interval.

Impact of Canal Water on Crop Yield

Agricultural productivity is a crucial factor in achieving higher growth in production, revenue and hence more returns. Yield depends on many factors, among which water availability is an important component. Considering this fact we did a trend analysis for pre-and post-canal period for the selected three crops.

Figure 5 shows the trend in yield for all three crops in three districts during the study period. The trend line indicates that after canal water, yields of oilseeds and pulses have increased sharply for all the three districts, while those of wheat have not changed much. In the post-canal period, Gandhinagar experienced higher mean yield for oilseeds and pulses than for wheat. On the other hand, Vadodara and Ahmedabad gained higher growth in yield for all three crops. Maximum growth in oilseeds is observed in Ahmedabad (39%) followed by Vadodara (35%) and Gandhinagar (20%). Ahmedabad and Vadodara both recorded 6 per cent growth rate in wheat (Table 1). Vadodara (54%) experienced higher growth in pulses followed by Ahmedabad (37%) and Gandhinagar (12%).

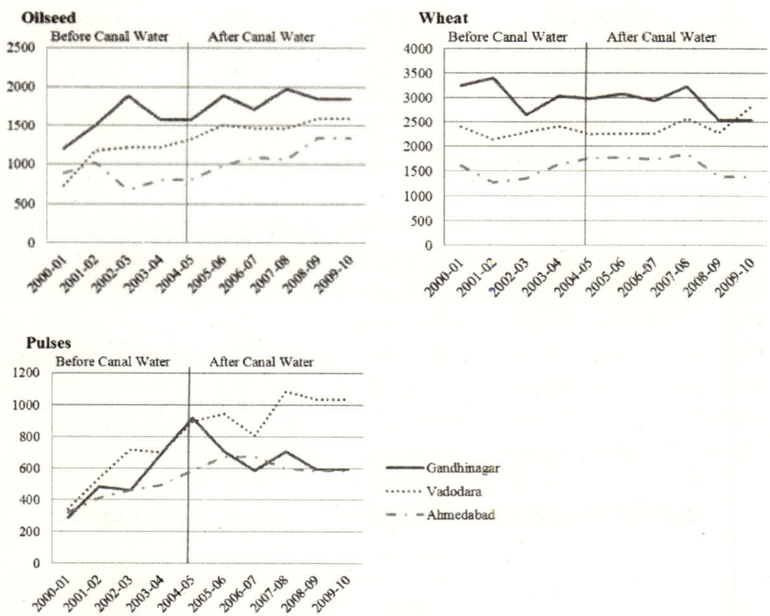

Figure 5: Trend in Yield (kg/ha) in Pre and Post Canal Period

Analysis of growth rate, mean and standard deviation for pre- and post-canal period for selected three crops (Table 3) shows that in most cases yield variability of crops has declined in post-canal period, due to timely availability of canal water. It also represents higher mean yield and higher growth rate in yield during post-canal period.

Table 3: Mean, SD, and Growth Rate of Yield for Selected Crops in Three Districts

District		*Gandhinagar*	*Vadodara*	*Ahmedabad*
Wheat	Mean–Pre	3058	2295	1527
	(SD)	(283)	(112)	(203)
	Mean–Post	2866	2435	1621
	(SD)	(318)	(258)	(221)
	Growth Rate (%)	6	6	6
Oilseeds	Mean–Pre	1543	1125	836
	(SD)	(246)	(231)	(129)
	Mean–Post	1853	1522	1161
	(SD)	(95)	(64)	(161)
	Growth Rate (%)	20	35	39
Pulse	Mean–Pre	568	638	454
	(SD)	(244)	(210)	(99)
	Mean–Post	638	982	624
	(SD)	(62)	(110)	(47)
	Growth Rate (%)	12	54	37

Note: Mean yield and SD in kilogram per hectare.

The summary results of significant increase in yield for all three crops are reported in Table 4. The *t* values of mean yield for pulses indicate that average yield increase in the post-canal period is statistically significant. Though wheat and oilseeds also recorded a fairly good average increase, it is not statistically significant. Standard deviation in the following table shows that during the post-canal-period the three districts experienced more stable yield compared to the pre-canal period owing to timely availability of water.

Table 4: Independent Sample *t* Test for Yield (kg/ha)

	Dummy	*N*	*Mean*	*Std. Deviation*	*t-value*
Wheat	0	5	6880	372	0.89
	1	5	6922	534	
Oilseed	0	5	3659	592	0.07
	1	5	4320	398	
Pulse	0	5	1661	540	0.05*
	1	5	2244	122	

* Significant at 95 per cent confidence Interval (CI).

Impact of Canal Water on Gross Revenue of the Crops

Gross revenue per hectare has increased by 345 per cent in Vadodara, 435 per cent in Ahmedabad and 310 per cent in Gandhinagar between the pre-and post-canal periods. However, the causality of impact of canal water on revenue is inconclusive. Gross revenue is influenced by changes in cropped area, in crop yield and in minimum support price (MSP). During the post-canal period, minimum MSP experienced large increases and hence higher gross revenue for farmers, which cannot be attributed to Narmada Canal water. Remunerative MSP for cotton, wheat and other crops announced by the central government may have provided strong incentives to farmers to increase production of these crops, provided farm inputs were sufficiently available (Shah *et al.*, 2009). Availability of relatively cheap and more certain Canal water could also have contributed to more cultivation of wheat. Although the cost of production has also gone up due to increasing fertiliser, pesticide and other input costs, the crop prices have increased much more, resulting in net revenue going up for these crops during the post-canal period.

Section 6 – Conclusions

It is evident from the above analysis that the availability of Narmada water has significantly increased the NSA and also average crop yield for wheat, oilseed and pulses in all the three districts that have so far received the water. The variability in yield and NSA has also declined, thus providing more stability to the farmers in planning their crops.

Another dimension is the current sumptuous water availability in the three districts studied owing to curtailed command area coverage at this point in time as compared to the full command area of the proposed canal irrigation system. These three districts are getting more water than what they would get once the command system is completely developed and made fully operational. This relative future paucity of water could create social stresses in the affected districts.

The study has a few limitations as well. Firstly, while drawing our conclusions, we could not rule out other confounders that could affect its outcome, like government policy, market environment, or incentives by government to increase production. Secondly, those factors that could have contributed to better agriculture in Gujarat, like improved market access are not captured in the present study. Ideally a matching control district not fed by any form of canal irrigation would have yielded stronger evidence of the impact of canal irrigation. However, inclusion of these three districts, being representative of canal irrigation, does provide enough evidence of the impacts of the Narmada Canal.

23

Irrigation Impacts of Sardar Sarovar Project Water: A Field Evaluation

Prerana Desai Vaidya

Background

Cultivation in Gujarat is dependent on the Southwest Monsoon and Kharif cropping predominates. The area sown and the timing are greatly influenced by the vagaries of the monsoon. Most of the state faces acute water shortage. The monsoon is erratic, resulting in drought conditions at the one extreme and floods at the other. A large part of the state receives low rainfall, characterised by large variation over time and space. Groundwater levels in many parts of the state have been declining. Overexploitation of aquifers has led to rising costs of irrigation from borewells and tubewells. Earning livelihoods from cattle rearing has also become very difficult owing to shortages of drinking water and fodder. Low literacy levels and the absence of technical knowledge severely limit alternative livelihood options in the rural areas, leading to increasing out-migration. In this context, the Sardar Sarovar project (SSP) is indeed a lifeline for the people of Gujarat. The study reported here examines the impact of water from SSP on the selected villages.

Selection of Regions/Villages

To assess the impact of SSP water it is necessary to study each region in its particular context. For the present study, the selection of villages was based on their location in the command area of the Narmada Main Canal, which has reached Kadi in Gujarat and the distributary network is ready up to Region 5. Region 1 was divided into two parts. From sub-region R_1P_1, five villages were chosen—Devaliya and Vagheli in Narmada District and Khodiya, Pala and Vadikua in Vadodara District. From sub-region R_1P_2, which has good soil, the five villages selected were Limdi and Dhantej in Vadodara District and Khandoli, Sama and Jantral in Panchmahal District.

Two villages were identified in Region R_2, Sinor-Vadodara, known for its fertile land. These are Valan in Vadodara District and Sarbhan in Amod Taluka of Bharuch District. Two villages covered in Region 4 (R_4), known as the Bara tract, are Samoj and Kangam in Bharuch District, both 8-10 km. from the Gulf of Khambhat coast.

The remaining ten villages were selected in Kheda District, Region 5 (R_5). According to the Sardar Sarovar Development plan this region is the most developed part of the command area. It ranks at no. 1 in respect of population, irrigation, crop intensity and index of value of production per ha of net area sown.

No villages were taken from Region R_3 as its characteristics are similar to the normal/good soil conditions of Bharuch District in Region R_2.

Table 1: Salient Features of Selected Regions for the Study

S. No.	*District*	*Talukas*	*Features*	*No. of villages sampled*
R_1P_1	Narmada-Vadodara	Tilakwada-Naswadi-	Hard rock; predominantly tribal area	5
R_1P_2	Vadodara-Pachmahal	Savali-Kalol	Medium black and brown soil suitable for groundwater development	5
R_2	Vadodara-Bharuch	Karjan-Amod	High agri. development with cotton growing area	2
R_4	Bharuch	Jambusar	Fertile land, saline groundwater with acute drinking water problems	2
R_5	Kheda	Kathlal-Kapadvanj-Thasra	Sandy soil and high irrigation intensity	10

A sample household survey was done in these villages. The parameters of the questionnaires of the survey were decided on the basis of an earlier study by the Narmada Planning Group (SSNNL, 1989).

The present study seeks examine the changes that were planned and expected to be brought about by SSP water in the lives of people. Water for irrigation became available to these villages from the year 2002 onwards. Region R_1 was the first to receive it. In Region R_5, where the distributor network was yet to be completed, a partial study was done of the irrigated villages. Regions R_2, R_4, and selected villages of R_5 had started receiving irrigation water by the end of 2004.

Methodology and Data Collection

As mentioned above, the findings of the project are based on a study of 24 villages in five districts—Narmada, Vadodara, Bharuch, Panchmahal and Kheda. The villages were selected on the basis of their location in the Narmada Main Canal command area. As this area is diversified, it has been divided into 13 regions based on their soil, climate, rainfall, level and variation, groundwater and temperature regimes. From each of the 24 villages approximately 50 respondents were selected randomly. Data was obtained from all the 50 through questionnaires administered in individual interviews. In addition to these, 25 persons such as the sarpanch, prominent farmers, president and secretary of the Pani Samiti (Water Users' Association—WUA) were interviewed. Focus group discussions were held with farmers and farm labourers. Each group had 8-10 people. The responses were manually recorded. A team of three to five people including the principal investigator and the field investigators visited each of the villages.

Analysis Plan

To assess the impact of SSP water it was essential to understand its distribution system. Irrigation water in the command area is delivered to farmers' groups, not to individual farmers. The farmers' groups have to manage distribution. It was further found that the system of Varabandhi (rotational irrigation) is practised for timely, assured and equitable water supply. It was most important to study the WUA in

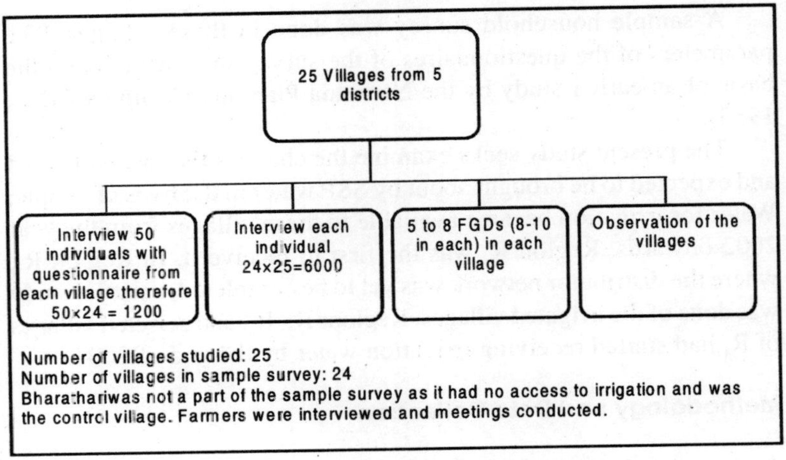

Sample Distribution and Methodologies Used

each selected village. Meetings were organised with the WUA members to discuss their attitude and approach towards the SSP system. The beneficiaries were asked if they found the irrigation system suitable and whether it helped them achieve higher agricultural output and income. They were also asked about any problems they faced, such as waterlogging, drainage and the steps taken to manage them. This is a highly technical and controlled distribution network. Therefore, they were asked how they had adapted to the canal irrigation system. Inquiries were made about their awareness and use of systems like drip irrigation and sprinkle irrigation in order to minimise water use and maintain soil quality by preventing waterlogging and salinity. The sarpanchs of all the villages were also interviewed in the context of their planning for improving village infrastructure like ponds and tanks, as now they had access to a water resource. Sample surveys, using questionnaires, were done to collect data from the beneficiaries of SSP Water. The questions asked covered family size, land holding, soil type, and crop yield and cropping pattern before and after the arrival of canal waters.

Background Characteristics of the Respondents

Background characteristics of the respondents, like gender, age, education, religion, etc. are shown in Tables 1 and 2. It can be observed that in regions R_1 and in R_5 male and female respondents are more or less in equal proportion; while in region R_2 and in R_5 more than 84 per cent of the respondents were male.

Table 2: Gender of the Respondents (%)

Gender	R_1P_1	R_1P_2	R_2	R_4	R_5
Male	53.5	53.5	95.0	84.2	51.4
Female	46.5	46.5	5.0	15.8	48.6

Table 3: Educational Characteristics (%)

Level	R_1P_1	R_1P_2	R_2	R_4	R_5
Pre-primary	16.8	20.4	12.7	11.3	7.2
1-4 std.	16.0	16.7	19.9	16.7	24.4
5-7 std.	21.3	21.8	24.3	13.6	31.2
8-10 std.	17.2	24.0	13.5	20.5	24.1
11-12 std.	4.5	12.5	16.9	19.2	8.6
12 std. plus	0.6	4.6	12.7	18.7	4.5

Table 3 shows the educational background of the respondents' family members. As primary education is compulsory, all children are enrolled. But the differences are sharp at the secondary and higher secondary levels. R_2 and R_4 have higher levels of education. Literacy rates according to the census of 2001 were 77.5 per cent for males and 62.3 per cent for females in R_2, and 92.9 for males and 58.5 per cent for females in R4. Literacy in the tribal villages of Naswadi and Tilakwada talukas in R1P1 is relatively low.

Table 4 shows agriculture as the main occupation of people in the selected villages. Combining percentages in the Farmer and Labourer categories, we found more than 75 per cent respondents working in agriculture. In the Housewife category, the percentage in R_2 and in R_4 is higher than that in R_1P_1, R_1P_2 and R_5. In Section 'Gender issues and SSP Water', we have indicated that census

Table 4: Occupational Characteristics of Respondent's Family Members (%)

Occupation	R_1P_1	R_1P_2	R_2	R_4	R_5
Farmer	46.7	34.9	39.6	34.7	61.2
Labourer	38.7	25.4	16.4	12.2	20.8
Housewife	4.7	7.1	21.2	20.2	3.0
Service	0.7	7.5	0.3	14.8	1.9
Business	0	1.3	2.5	0.3	0
Social work	0	6.4	0	0	1.1
Study	9.3	13.2	20.0	17.3	21.6
Tailor	0	0.4	0.5	0.5	0
Hawker	0	0.9	0	0	0.4
Hotel	0	0	1.1	0	0
Driver	0	9.2	0.6	0	0

figures for 2001 reveal the percentage of female agricultural labour also to be higher in these three regions. It may be noted here that socio-economic backwardness combined with low literacy seems to leave few work options other than marginal farming or agricultural labour.

The Service category, which includes employment in the organised sector and in non-farm activities as wage labour, the percentage is larger in R_4 than in the remaining regions. Literacy rates of both males and females are also higher in this region. We have already pointed out that the villages in R_1P_2 rank second in terms of percentage in the Service category. Relatives of the respondents work in the surrounding industries and as Drivers. About 2.5 per cent family members of the respondents in R_2 are in the Business category, which is the highest among all the regions.

Table 5: Religion of the Respondents (%)

Religion	R_1P_1	R_1P_2	R_2	R_4	R_5
Hindu	81.6	87.5	17.9	100	98.6
Muslim	10.3	12.5	82.1	0	0
Christian	0	0	0	0	1.4
Kabirpanth	8.1	–	–	–	–

Table 5 shows the religious background of the respondents. Each village has the predominance of people a particular religion. Of the five regions, Hindus predominate in four and Muslims in one. In R_4 all respondents were Hindu Rajputs of the Padhiyar or Sindha communities. In R_2, 82.1 per cent were Muslim and 17.9 per cent were Hindu. In R_5, 98.6 per cent respondents were Hindu and remaining 1.4 per cent was Christian.

Table 6: Broad Caste Classification of Respondents (%)

Caste	R_1P_1	R_1P_2	R_2	R_4	R_5
SC	25.3	14.1	5.1	4.0	18.6
ST.	58.6	1.6	8.7	17.0	4.3
OBC	9.2	56.3	7.6	66.6	24.4
Others	6.9	26.6	78.6	12.4	55.7

From the data in Table 6 we find that in region R_1P_1 25.6 per cent of the respondents were SC and 58.6 per cent ST. Thus R_1P_1 is predominantly inhabited by vulnerable social groups. In R_1P_2, 56.3 per cent were Baria Rajput, classified as OBC, 14.1 per cent SC and only 1.6 per cent were ST. In this region 26.6 per cent were from groups other than SC, ST and OBC. They were Rajput. Thus, in the R_1P_2 region a majority of inhabitants were Rajput.

In region R_2, 78.6 per cent of the respondents were Muslims of the Bohra community, while 8.7 per cent were from the SC category. In region R_4, in selected villages of Jambusar Taluka, all the respondents in one village were Padhiyar Rajput, an OBC. In another village they were a mix of 4 per cent SC, 17 per cent ST and 12.4 per cent higher caste Rajput.

In the R_5 region, 18.6 per cent respondents belonged to the SC group, 4.3 per cent to the ST group, 24.4 per cent were OBC and the remaining 55.7 per cent included Parmar community and Patidars.

Table 7: Housing Characteristics of Respondents (%)

Type	R_1P_1	R_1P_2	R_2	R_4	R_5
Kuchcha	63.2	7.8	18.2	2.0	2.6
Pucca walls/tin roof	25.2	57.9	45.9	61.4	52.9
Pucca	11.5	34.4	35.9	36.6	44.5

Table 7 shows that 88.4 per cent of the respondents in R_1P_1 lived in kachcha or semi-pucca houses. In the remaining regions the majority lived in semi-pucca houses. In R_2, which is economically more developed, only 18.2 per cent live in kachcha houses. These are migrant labourers belonging to the SC or ST groups. The condition of these kachcha houses is superior to that of the 63.2 per cent respondents in R_1 living in kachcha houses. Except in R_1P_1 34-44 per cent of the respondents live in pucca houses. Domestic power supply is available to more than 90 per cent of the respondents. However, many who are agricultural labourers living on the outskirts of villages, do not have power.

Table 8: Land Owned by Respondents (%)

Vighas	R_1P_1	R_1P_2	R_2	R_4	R_5
1-20	93.1	82.8	74.4	61.0	87.2
20-40	4.6	9.4	15.4	32.9	7.1
41-70	1.1	1.6	10.3	6.1	5.7

Table 8, shows that of the respondent land owners in the selected villages, 93.1 per cent in R_1P_1, 82.8 per cent in R_1P_2, 74.4 per cent in R_2, 61.0 per cent in R4 and 87.2 per cent in R5 are small and marginal farmers. In regions R2 and R4 25-39 per cent of land owners own 20-70 bighas.

Table 9: Source of Income of the Respondents' Families (%)

Source	R_1P_1	R_1P_2	R_2	R_4	R_5
Farming	51.2	55.2	46.5	60.0	52.3
Farm labour	36.7	19.0	17.9	19.5	34.8
Service	2.4	14.7	11.9	18.3	2.3
Self-employed	0.6	3.4	9.5	12.2	8.0
Milk production	0	6.9	7.1	3.0	9.1
Other	9.0	8.9	7.1	0	.08

Note: An individual could be a farmer as well as farm labourer or in service sector leading to duplication of response and a higher percentage total.

Table 9 shows the income sources of respondents in the sampled area. For the present study only agro-based villages were selected.

Obviously the major source of income is agriculture. It is evident that the respondents earn their livelihoods through cultivation as well as farm labour. Small and marginal land holders also work as labourers in their own villages or outside. When confronted by vagaries of weather, take recourse to whatever other livelihood options are available.

Income from Labour in R_1P_1 was 36.7 per cent, whereas in R_2 and R_4, at 17.9 per cent and 19.5 per cent respectively from the same source. Canal water has created tremendous work opportunities both for cultivators and labourers to generate more and more income for the betterment of their lives.

Region-wise Post SSP Crop Cultivation

Table 10 shows the impact of SSP water on the winter cropping pattern in the selected villages. Canal irrigation has made cultivation of Rabi crops possible in all the surveyed villages. Of the respondents, 22.1 per cent in R_1P_1, 80.9 per cent in R_1P_1, 62.5 per cent in R_2, 53.3 per cent in R_4 and 98.0 per cent in R_5 cultivated wheat in the winter of 2007. Thus wheat has emerged as the most prominent crop in the Rabi season. Maize being the staple food in R_1P_1, 75.7 per cent cultivated it with the help of canal water. Jowar is sown late in the monsoon. The fodder crops rachko and sundhiya are commercially cultivated in R_2 and 10 per cent cultivate rachko only for sending outside their district. Fennel seed (variyali) was introduced for the

Table 10: Region-wise Crop Cultivation in Winter after Canal*(%)

Crops	R_1P_1	R_1P_2	R_2	R_4	R_5
Wheat	22.1	80.9	62.5	53.3	98.0
Maize	75.7	29.1	8.3	0.7	0
Pulses	20.8	10.9		23.2	9.3
Sugarcane	0	0	8.3	0	0
Jowar	12.7	2.3	42.1	34.4	2.0
Fennel seed	0	0	0	0	2.0
Sundhiya (Grass)	0	0	10.7	0	9.0
Rachko (Grass)	0	0	10.7	0	8.0

* The above table has multiple responses, hence the total does not add to 100.

first time in some of the surveyed villages in R_5. In Rampura village 2 per cent respondents successfully experimented with this new crop.

Table 11: Summer Crops* (%)

Crops	R_1P_1	R_1P_2	R_2	R_4	R_5
Bajra	13.0	63.3	42.9	81.6	86.7
Maize	37.0	22.4	0	1.1	33.3
Vegetable	24.8	28.2	28.6	35.3	45.7
Sundhiya	9.0	17.2	39.2	27.8	41.2
Fodder Vegetables	4.0	28.2	21.2	19.6	27.5

* The above table has multiple responses, hence the total does not add to 100.

Table 12: Monsoon Crops* (%)

Sr. No.	*Crops*	R_1P_1	R_1P_2	R_2	R_4	R_5
1.	Cotton/Tuver	40.8	26.3	30.7	35.3	34.1
2.	Tobacco	0	32.7	0	0	15.5
3.	Paddy	5.0	20.6	15.3	10.3	68.5
4.	Bajra/Tuver	0	25.7	0	0	0
5.	Maize/Tuver	25.0	27.6	3.4	0	2.7
6.	Pulses	30.0	0.5	35.3	37.6	9.3
7.	Sesame (Til)	0	0	6.8	13.2	0.9
8.	Paddy (non-irrigated)	7.5	5.3	0	1.1	0
9.	Castor	3.3	13.9	0	0.3	4.1
10.	Sugarcane	0	0	0	0	0
11.	Vegetables	0	1.3	10.2	1.6	11.7
12.	Sundhiya (fodder)	0	0.8	13.4	0.5	1.8
13.	Jowar	0	0	9.8	25.9	0
14.	Foundation crops-Cotton, Bajra etc.	0	10.0	0	0	0

* The above table has multiple responses, hence the total does not add to 100.

Tobacco was the most preferred crop in R_1P_2, R_1P_1 and R_5 until the SSP water arrived. Cultivators in R_1P_2, such as village Limdi in Savali Taluka and in the villages of Kalol Taluka, have started cultivating alternative crops like cotton and cotton seeds. Big farmers in R_1P_2 preferred to go for foundation crops of cotton, bajra, maize, etc. Jowar, castor, grass (sundhiya and rachko), vegetables and foundation crops are taken up in the months of August-September.

Of the respondents, 25.9 per cent in R_4 and 9.8 per cent in R_2 cultivate jowar, a staple food for many in Bharuch and Surat Districts.

Impact on Agriculture

Irrigated Rabi crops are now possible in all the surveyed villages in addition to the Kharif crop. Wheat has emerged as the most prominent Rabi crop. Maize, a staple in both the parts of R_1, is another favoured item with the advent of canal water. Conditions now exist for optimum cropping pattern and higher cropping intensity. This, along with higher productivity, holds promise of a rise in as well stability of income.

Table 13: R_1P_1: Agricultural Profile before Canal

	Kharif	*Rabi*	*Rate/ql (Rs)*	*Production/bigha (ql)*	*Income (Rs.)*
Black Soil	Bt Cotton	–	1600	4 to 5	7,500
	Paddy	–	800	2 to 3	2,400
Brown	Pulses	–	3000	1 to 2	3,200
	Maize	–	750	2 to 3	1,500

Table 14: R_1P_1: Impact of SSP Water on Agricultural Yield and Income

Soil Type	*Kharif*	*Rabi*	*Rate/ql (Rs.)*	*Production/bigha*	*Income (Rs.)*
Black Soil	Bt Cotton	–	2250	8 to 9	20,250
	Paddy	Wheat	800	8 to 9	7,200
			1100	6 to 7	7,700
Medium Black and Brown	Maize	–	900	5 to 6	3,600
	Pulses	–	2500	3 to 4	10,000
	Tuver	–	2500	2 to 3	7,500
	Tobacco	–	2000	5 to 4	8,000

Tables 13 and 14 show the agriculture situation of region R_1P_1 before and after the canal. In the villages of Tilakwada and Naswadi talukas with a predominantly tribal population, where the land holdings are marginal, SSP water for irrigation became available from the year 2002 onwards. About 80 per cent farmers of these villages now completely avail of the facility and have started taking three to five crops a year. Crop production has increased by 50-100 per cent, and the income of farmers holding 4-5 bighas has increased three to four times after the canal became operational. The production of food grains and pulses has improved their lives, as they now have enough food. Rabi production yields good returns. Cotton sown in June-July requires nine months. Tuver is intercropped with cotton. Many farmers also cultivate other pulses.

Region R_1P_2

Irrigation in this region started in the year 2002 through minors. Farmers now take crops in two seasons. Tables 15 and 16 show the impact of SSP water. A major gain is Rabi cultivation, which has increased production and incomes. Irrespective of land holdings, farmers now cultivate wheat. Big farmers are experimenting with new crops like cotton Bt cotton and sunflower, and cultivating foundation seeds for various crops, which they supply to nearby companies. Ultimately these activities will encourage small and marginal farmers to change cropping patterns. The impact is notable in the villages of Savali and Kalol talukas, where entrepreneur farmers have started investing in the farms of small and marginal farmers, leading to stable and higher incomes for small and marginal farmers and profits for the entrepreneurs.

Table 16 shows that during the Kharif season production depended entirely on the vagaries of weather. Until now 80 per cent of the total cultivable land was under tobacco. Bajra, maize, paddy and tuver were cultivated only for home consumption. Owing to heavy

Table 15: R_1P_2: Agriculture before Canal

Soil type	*Kharif*	*Rabi*	*Production (ql/bigha)*	*Income (Rs.)*
	Crop	*Crop*		
Black	Paddy	Nil	2 to 3	2400
	Castorseed		1 to 2	600
Brown and Mixed	Tobacco	Nil	2 to 3	7500
	Bajra	Nil	2 to 3	1800
	Maize	Nil	2 to 3	1800
	Tuver	Nil	1 to 15	For home consumption

Table 16: R_1P_2: Cropping Pattern after Canal

Soil Type	*Kharif*	*Rabi*	*Prod'n/bigha (ql)*	*Income/bigha (Rs.)*
Black	Cotton		10	18,000
	Bt Cotton		10 to 15	33,000
	Paddy			10,400
	Castorseed		8 to 10	8,000
Brown and Mix	Tobacco		5 to 6	12,000
	Tuver		For personal use	
	Bajra	Wheat Maize	10 to 12	9,600
	Maize		15 to 20	16,000
	Paddy		10 to 13	13,000
Foundation Crop	Bajra Tuver Maize Pulses Cotton Bt Castor Seeds		10 to 12 13 to 15 8 to 9 10 to 15 10 to 12	

rains over the last three years, the production of maize badly suffered all over Gujarat. In this area, one of the major advantages of SSP water was that it made possible Rabi cultivation, which has increased production as well as incomes. The input cost for water has dropped from Rs. 1500-1600 to Rs. 240-300 per crop.

Region R_2

Table 17: R_2: Agriculture in Valan and Sarbhan villages

Soil Type	*Kharif*	*Rabi*	*Prod'n/bigha (ql)*	*Income (Rs.)*
Black Soil	Cotton		6 to 7	14,700
	Pulses Tuver		3 to 4	10,000-15,000
	Rachko (grass)	Rachko (grass)		

SSP water reached this region in 2003-04 and wheat was sown for the first time. Small and marginal farmers, no longer dependent on borewells, now sow Rabi crops. Canal water has reduced electricity costs for borewells. Cotton needs two rounds of watering after the

Table 18: R_2: Impact on Cropping Pattern

Soil Type	*Kharif*	*Rabi*	*Prod'n/bigha (ql)*	*Income (Rs.)*
	Cotton		10 to 12	25,200
Black soil	Pulses			
		Wheat	10 to 12	14,400
	Rachko (grass)			

monsoon. If this need is met, production increases 50 per cent-100 per cent. According to the beneficiaries, canal water protects crops when rainfall is poor and enhances production and income. With SSP water, beneficiaries in these two villages are marching to prosperity.

Region R_4

Table 19: R_4: Agriculture before Canal

Soil Type	*Kharif*	*Rabi*	*Prod'n/bigha (ql)*	*Rate/ql (Rs.)*	*Income (Rs.)*
Black Soil	Cotton		4 to 5	1600	8,000
	Digvijay Jowar		4 to 5	700	3,500
Brown Soil	Bajra		3 to 4	800	3,200
	Tur		1 to 2	2500	5,000
	Sesam		1 to 2	3000	6,000
	Pulses		2 to 3	2800	8,400

Table 20: R_4: Cropping pattern after irrigation

Soil Type	*Kharif*	*Rabi*	*Prod'n/bigha (ql)*	*Rate/ql (Rs.)*	*Income (Rs.)*
Black Soil	Bt Cotton		4*	2200	8800
	Jowar		10 to 12	800	9600
	Tur		5 to 6	3000	18000
Brown Soil	Bajra		8 to 9	800	7200
	Pulses		5 to 6	3200	19000
	Seasam		4 to 5	5000	2500
		Wheat	10 to 12		

* Bt Cotton production decreased in 2006-07 due to virus infection of the crop, as per information given by the farmers.

These villages of the Bara region were selected because of their proximity to the Gulf of Khambhat coast. SSP Water reached here in 2003 and 80 per cent of the area is now irrigated. In planning for the command area, a special irrigation strategy was adopted here. The crop which used to fail due to poor rain, especially during the last round of monsoon, now gets enough water and the farmers are no longer apprehensive. They are now aiming at outputs on par with Sarbhan and Valan Villages in R_2.

Region R_5

This was the most developed region in every aspect of agricultural requirements. Paddy, wheat and tobacco are the important crops. Tobacco is cultivated on some fairly large, well drained pockets of sandy loam. Ten villages of the region were selected for study, but they revealed wide diversity in soil conditions, groundwater levels and availability of drinking water. The villages Kathana, Rampuralat and Laxmanpura have no groundwater source. Bharathai is the only village in the region that has not received any canal irrigation.

SSP water was made available for irrigation in R_5 from 2003-04 onwards. The infrastructure required for agricultural development was superior to the rest of the region of the Narmada Main Canal. Overexploitation of groundwater in this area had led to scarcity and salinity in groundwater, but within two years cultivators started reaping the benefits of SSP. R_5 has villages with no irrigation, partial irrigation, and canal irrigation. So agriculture in these villages provides an opportunity to compare between them and identify the benefits of canal irrigation.

The paddy growing farmers of Dujewar and Navapura in Kheda region informed us that the yield in a normal monsoon before the pre-canal days was 8-10 quintals per bigha, which increased in 2006 to 15-16 quintals per bigha after the canal. The requirement of paddy of irrigation at regular intervals and in necessary quantity was fulfilled by canal irrigation, resulting in doubling of production. Beneficiaries of Ranpuralat village experimented successfully with a new crop—fennel seeds—during the winter of 2007. This fetched them very good returns. This may be seen as an example of encouraging entrepreneurial spirit among the farmers, made possible by canal irrigation.

Table 21: R_5: Agricultural Profile of Selected Villages before and After Canal Water

Soil Type	*Village name*	*Crop*	*Yield per bigha (ql)*		*Av. income per vigha (Rs.)*	
			Before Canal	*After canal*	*Before canal*	*After canal*
Black	Bharathari	Paddy unirrigated	6 to 7		6,500	
Brown		Cotton	4 to 5		10,000	
		Tuver	2 to 3		7,500	
Black and Sandy Loam	Kathana Rampuralat Laxmanpura	Wheat on residual moisture	1 to 2		1,600	
Saline		Cotton	4 to 5		8,000	
	Anara	Tobacco	4 to 5	8 to 9	15,000	27,000
		Paddy	1 to 1.5		1,200	
Black	Dujevar	Paddy irrigated		15 to 16		16,000
	Navpura	Castor seed		3 to 4		20,000
	Navpura	Bajra		10 to 12		11,400
		Maize		6 to 7		8,000
		Tuver		4 to 5		10,000
		Sesame		3 to 4		15,000
		Wheat irrigated		10 to 12		11,400
		Fennel seeds		3 to 4		20,000

Table 22: Change in Income Levels with Development of Irrigation (Rs.)

Crops	*Income with no irrigation*	*Partial irrigation before canal*	*Impact of SSP water after canal*
Paddy	6,500	10,000	16,000
Cotton	10,000	8,000	12,000
Tuver	7,500	10,000	10,000
Wheat	Nil	6,400	14,400
Cotton	Nil	5,000	11,400
Bajra	Nil	10,000	6,400

Note: Annex Table A.1 gives Price of Different Crops Considered for the Assessment of Income

Table 22 shows the change in levels of income for different crops with the advent of irrigation, considering agricultural output and water availability. It is evident that income and yield per unit area have increased with SSP water becoming available.

Impact on Work Opportunity and Wage Structure

Thus Farmers are opting for at least one commercial crop a year and use part of their land for cultivating produce for home consumption. Moreover they are able to save the Kharif crop if one of the rounds of the monsoon fails and also to grow the Rabi crop. Intensive and successful farming activity throughout the year leads to greater and more stable income. This is the greatest opportunity. It creates a new intensive work timetable for farmers in all the regions and hence creates income. SSP water has benefited small and marginal farmers tremendously and made them independent financially. They are able to cultivate without taking heavy loans from big farmers and paying them a heavy price for borewell water. As a result there is an acute shortage of and demand for farm labourers in all the regions. A specific time table is followed in agricultural practice according to the crop sown. The requirement of labour fluctuates accordingly.

SSP water has created tremendous work opportunities for both the farmers and farm labourers. In $R_1P_{1,}$ more than 60 per cent farmers had 11 to 20 days of employment before the advent of the canal. After canal, 90 per cent of them have work for more than 20 to 25 days. In $R_1P_{2,}$ 32.8 per cent respondents have employment up to 20 days in a month and more than 60 per cent get work through the month.

In $R_{2,}$ Sinor Vadodara, agricultural activities did not come to a halt because there was adequate groundwater. After the canal, the introduction of irrigated wheat and use of irrigation in cotton production with 2 to 3 rounds of watering, employment increased by 24.5 per cent with up to twenty days per month and 75.5 per cent percent of the respondents have productive work through the month.

Region $R_{4,}$ the Bara tract comprising alluvial soil, used to suffer owing to uncertainty of rains. Farm employment was limited to the monsoon. Increase in the cotton yield and other crops like jowar,

Table 23: Farm Workload before and after Canal (%)

Farm work load (no. of days)	*R_1P_1*		*R_1P_2*		*R_2*		*R_4*		*R_5*	
	Before	*After*	*Before*	*After*	*Before*	*After*	*Before*	*After*	*Before*	*After*
	23.0	0	26.6	3.1	12.8	0	1.2	2.9	2.9	0
	62.0	1-10	34.6	32.8	60.8	24.5	75.6	60.0	34.3	0
	12.2	11-20	17.2	64.1	25.4	75.5	0	100.0	60.0	17.1
No response	10.3	21-30	0	0	0	0	0	0	0	0

pulses, sesame and irrigated wheat crops have created sufficient productive employment for the entire population of the region. Where 60 per cent respondents had work for less than 20 days a month before the advent of canal water, 100 per cent percent now get to work.

R_5 was the most developed of the thirteen regions when planning for SSP was done. It was due to overexploitation of groundwater that crops could not be irrigated when they most required it. This is of particular concern in paddy cultivation. The last round of the monsoon is a must to ensure both quantity and quality of paddy yield. Now 82.9 per cent respondents in R_5 have employment for all seasons. Crop like irrigated wheat and fennel seeds are cultivated after canal water became available.

Table 24: Change in Daily Wages after Canal (%)

	R_1P_1	R_1P_2	R_2	R_4	R_5
Yes	86.2	70.3	46.2	80.5	82.9
No	1.1	9.4	10.3	9.8	15.7
No Response	12.6	20.3	43.6	9.8	1.4

Table 25: Pattern of Daily Wages after Canal (%)

In Rs.	R_1P_1	R_1P_2	R_2	R_4	R_5
20-40	11.5	60.9	25.6	63.4	97.1
41-60	71.3	0	26.5	0	1.4
61-90	1.1	0	0	0	1.4
No Response	16.1	39.1	53	36.6	0

The impact of SSP water on work opportunity for farm labourers in the regions is presented Tables 24 and 25. The region-wise outcomes are discussed below:

Region R_1P_1: After SSP, there is an acute shortage of farm labourers in this area. The requirement of labourers has increased with the rise in farm activities. The wage of farm labourer has also increased by 40-50 per cent. Each crop has a particular time table. Maintaining this time table is very important. When the demand for labourers is high the daily wage rates also rise.

Region R_1P_2: Agricultural activities have increased after the SSP water reached the region. This created tremendous work opportunities for farmers such as cultivating all their land, protecting the Kharif crop when one of the rounds of the monsoon fails, and cultivating a Rabi crop. This presents tremendous opportunities and a new working timetable for farmers in the region. Small and marginal farmers are availing themselves of the benefits of SSP water. They are can cultivate without becoming financially obliged to big farmers.

Region R_2: A rise in the yield of cotton due to the SSP water has generated more work opportunities in the region. The work load of farmers has also increased owing to the cultivation of an additional crop. Canal water has made cultivation possible for marginal and small farmers. Labourers who work in cotton fields during harvesting time earn a minimum of Rs. 250-300 per day. With the expansion of farming activity through the year, farm labourers earn about 50 per cent more from commercial crops.

Region R_4: Water availability for irrigation has vastly increased work opportunities for farmers. Rabi season cropping has brought in new activities and a new time table.

Region R_5: With all its inherent pre-canal advantages (and also disadvantages due to overexploitation of groundwater and salinity ingress) described earlier, SSP presents this region with a whole new range of work opportunities for both cultivators and farm labourers. The socio-economic condition of the people here is far superior to that of villages in the nearby Naswadi Taluka or villages of Panchmahal District.

Impact on Groundwater and Drinking Water

Table 26: Source of Drinking Water (%)

Source	R_1P_1	R_1P_2	R_2	R_4	R_5
Tap	18.8	53.2	76.9	69.4	75.8
Hand Pump	15.6	36.2	23.1	0	9.2
Well	3.2	2.9	0	30.6	14.5
River	34.5	0	0	0	0
Pond	0	0	0	0	4.0

Table 27: Types of Water Facility for Farming in Percentage (%)

Type	R_1P_1	R_1P_2	R_2	R_4	R_5	*Total*
Canal	66.0	68.8	49.8	77.4	75.7	68.2
Well	4.0	20.0	24.6	21.7	0	13.7
Tubewell	2.0	0	17.0	0	0	1.4
Pond	0	0	0	0.9	5.4	1.2
River	7.0	0	0	0	0	1.7
No Response	21.0	11.3	8.6	0	18.9	

The impact of SSP on the groundwater and drinking water situation are given in tables 26 and 27. The region-wise situation is described below.

Region R_1P_1: SSP has served in this region to recharge groundwater, which fulfils drinking water needs in this region. The water table now is at 100-150 feet. Borewells with hand pumps exist in all the villages.

Region R_1P_2: Most dug wells here had gone dry. Only some big farmers could afford borewells. The Narmada canal has served to recharge groundwater and natural water courses. The dry wells, especially those close to river banks, got recharged up to 30 to 40 feet. Thus the groundwater and drinking water situation has improved.

Region R_2: The situation of groundwater and drinking water was satisfactory in the selected villages of this region. Most cultivators use borewell irrigation for cotton as they think that the temperature of groundwater is more suitable for this particular crop. The farmers apprehend waterlogging in the future. They feel they need to learn more about management of canal irrigation, drip irrigation and sprinkle irrigation.

Region R_4: The potential of groundwater development is meagre in this region, which is under a special category for irrigation purposes. Surface soil is heavy-textured and saline. The regionalisation planning for SSP recommended the creation of new tanks and filling up of the existing ones. Groundwater levels have improved in the region, but it is saline in many wells. Farmers use water from irrigation ponds built by SSP.

Region R_5: The areas of Kathana, Rampuralat and Laxmipura villages have no groundwater. Hence they are 'no source' villages.

Drinking water supply in the selected villages is through pipelines and handpumps. At Dujevar and Navapura additional borewell-*cum*-hand pumps have been provided as supply from the water supply scheme is insufficient. Bharathari village has acute shortage of drinking and irrigation water. A borewell-*cum*-hand pump has provided in this village. About 8000 hectares area in Thasara Taluka has acute groundwater shortage. In 100 hectares area around Bharathari village the water is saline.

In Kathana village the groundwater level has risen by 8 to 10 feet but the water is not potable. Some portion of Ladwel village of Kathlal taluka has come under the adverse effect of SSP water as the Narmada main canal passes through the village territory. Some parts of fields have also become damp due to penetration of water.

Groundwater and drinking water has generally improved in all regions. However, impact differs according to groundwater and soil conditions in the various regions. Farmers in some areas apprehend waterlogging due to the SSP water and expressed need to learn more about the surface irrigation.

Impact on Animal Husbandry and Dairy Activity

Table 28: Arrangement of Fodder before Canal (%)

S. No.	*Source of Fodder*	R_1P_1	R_1P_2	R_2	R_4	R_5
1.	Was getting fodder	11.5	12	41.0	17.1	17.1
2.	Buying from outside	8.0	13	5.1	24.1	58.6
3.	Wasn't Keeping animals	44.5	47	37.0	48.8	15.7
4.	Drop animals in forest	36.0	38	16.9	20.0	8.6

Table 29: Arrangement of Fodder after Canal (%)

	R_1P_1	R_1P_2	R_2	R_4	R_5
1. Can get fodder	66.2	73.8	79.7	73.7	72.8
2. Buy from outside	0	0	0	0	1.5
3. Don't keep animals	33.8	26.2	20.3	26.3	27.2
4. Sell fodder	No	No	Yes	Yes	No

Table 30: SSP Water Impact on Expenditure on Animals in Surveyed Villages

Villages	*Region*	*Before canal (Rs./month)*	*After canal (Rs./month)*
Sankheda Savali	(R_1P_1)	1,550	900
Sankheda Savali	(R_1P_2)	4,052	2,961
Sinor Karjan	(R_2)	3,875	362
Bharuch Jambusar	(R_4)	2,367	1,500
Mehmdabad Daskroi	(R_5)	6,410	2,836
Total		4,146	2,315

Tables 28-30 show the impact of SSP water on dairy activity. Besides strengthening the farm sector, canal irrigation has given a boost to animal husbandry and dairying, generally associated with agriculture in Gujarat. Farmers while planning a crop pattern take into account the feed requirements of milch animals. With fodder easily available and reduced expenditure on water, the numbers of people rearing cattle and of animals they rear have increased.

Table 31: R_5: Animal Husbandry and Milk Production in Selected Villages

Village	*Cows*	*Buffaloes*	*Bullocks*	*Draught animals*	*Poultry*	*Milk yield (l./day)*	*Income (Rs./day)*
Kathana	50	500	100	500	100	800	16,000
Rampura		120	25	50	10		
Laxmanpura							
Dahiap							
Navapura	250	300	240	350	0	1100	22,000
Dujevar							
Ladwel	27	424	172	38	35	800	16,000
Lasundra	500	3000	700	300	100	1200	24,000
Anara	250	500	300	700	100	1200	24,000
Bharathari	70	300	250	Nil	Nil	800	16,000

Note: Purchase price of milk was assumed at Rs. 20 per litre. for the above calculation.

Table 31 provides data on animal husbandry and milk production in R_5 region (Kheda district). Surveys of the selected villages revealed farmers in Bharathari village who survive on the income from milk in the absence of adequate agricultural output. The minimum production of milk in every village is 800 l./day, which is supplied to the Amul Dairy. Lasundra has four milk cooperative societies. The Rabari and Bharvad communities in Lasundra and Anara generally found sell milk in the open market or to nearby towns. Milk compensates for low income from agriculture. Analysis of the survey findings shows that fodder shortage has just begun to be overcome with increase in the agricultural activities.

With the availability of SSP water fodder is now cultivated side by side with the main crop, on the edges of fields. The benefit of SSP water is that feed for milch cattle does not require any extra expenditure. Thus income from the milk would become a surplus for the farmers of these villages. Moreover, irrespective of the land held, farmers keep a pair of bullocks for farm activities. Hence, a great advantage of SSP is improved profits from animal husbandry. It was observed that consumption of milk and milk products has gradually disappeared from the diet of the rural people except in the Kheda region. It can only be hoped that the boost in agricultural and dairy activity will lead to improvement in the dietary habits of the people.

Impact on Gender Issues

The impact of irrigation on gender issues involves a wide range of factors. These include direct and immediate impact as well as indirect, secondary and long-term impact on socio-economic structures. The direct impact could be categorised in terms of elimination of the drudgery of drawing water making time available for alternative productive work and higher income. These factors in turn are related to the indirect impact of family size, sex ratio, literacy levels, treatment of women and their status in general.

Table 32: Family Members of Respondents Going Out of Village for Farm Labour (%)

	R_1P_1	R_1P_2	R_2	R_4	R_5
Yes	67.8	6.3	0	0	24.0
No	22.2	89.1	97.6	100.0	76.0
No Response	12.6	4.7	0	0	0

In the villages studied, no women had to carry water on their heads. All the villages have drinking water supply through different water supply schemes such as Panchayat water works. In some villages drinking water is supplied through state level schemes. Villages in Kheda District also receive drinking water from the District Water Supply scheme.

The main concern here is the high percentage of women working as agricultural labourers. With the availability of SSP water agricultural production has increased in all the villages. Due to migration of the men in certain areas, the thrust of agricultural activity is borne by the women. This has resulted in an increase in labour-intensive work which has a direct effect on the workload of women. Better production is resulting in buying more domestic animals. The animal husbandry sector too is primarily managed by women. Once again, therefore, increase in numbers of animals results in heavier workload for women. The major impact of availability of water, therefore, will result in increase of women's workload. Steps need to be taken in order to ensure care for the health and education of these women.

Impact on Taking Crop Loans

Stabilisation of agricultural activities, availability of ground and drinking water, improved profitability of animal husbandry and dairy activities all together brought about self-confidence among the farmers. Table 33 shows the percentage of respondents taking loans for crops before and after the advent of canal water. It is evident that in the latter situation, readiness to take advance crop loans has increased in all the regions except in Valan village of $R_{2.}$ The religion of the cultivators here prevents them from taking loans. Before the canal, 52.4 per cent in R_1P_1 used to take crop loans. After the canal this proportion is up to 61.9 per cent.

Table 33: Impact of SSP Water on Taking Crop Loans (%)

	R_1P_1		R_1P_2		R_2		R_4		R_5	
	Before	*After*	*Before*	*After*	*Before*	*After*	*Before*	*After*	*Before*	*After*
Yes	52.4	61.9	18.2	77.8	13.3	13.3	36.2	48.3	18.4	92.4
No	47.6	38.1	82.8	55.6	86.7	73.3	58.6	51.7	81.6	07.6

Region R_1P_2 has witnessed a lot of change in risk taking ability. About 78 per cent now take crop loans against only 18.2 per cent before the canal. Despite the religious taboos against taking loans, there is a decrease in the percentage of respondents saying 'No' to loans.

In R_4 respondents were not ready to take loans because the irrigation facility was slow to reach them. Now that it is assured, they are ready to take advance loans.

Respondents in R_5 are more aware of the advantages of irrigation. Only 18.4 per cent of them were taking crop loans before the canal. Now this percentage is up to a spectacular 92.4.

Their faith in themselves is considerably strengthened. In Kheda, 92 per cent said that they were now taking crop loans. In Panchmahal and Savali the proportion is 70 per cent. Gradually they have started feeling strongly about their capacity to build a good life for themselves and their families. They themselves see a marked difference in their living standards. A poor farmer of Jantral village said that he could now afford to have a bath and change his clothes at the end of day's labour.

Impact on Living Standard

Most respondents said that SSP water had a positive impact on their living standard (Table 34). The priority of farmers with enhanced income after the advent of the SSP water is renovation of their houses, from kachcha to pucca, semi-pucca to pucca and so on. Thus the overall aim is to improve their living conditions.

Table 34: Change in Living Standard after Canal (%)

Region	*Yes*	*Not much Difference*	*No Response*
R_1P_1	78.2	21.8	0
R_1P_2	92.2	3.1	4.7
R_2	63.4	0	36.6
R_4	91.5	0	8.5
R_5	98.6	0	1.4
Total	86.3	6	13.1

The second priority is to buy milch animals and bullocks or

increase the number of animals reared. They wish to join milk cooperatives and increase their income further.

Also in the list of the investment of small farmers would be agricultural equipment such as like tractor, thresher, diesel pump-set and purchasing a motorcycle or scooter. The respondents who did not have power connections were aspiring to them with the rise in incomes. Thus the basic purpose for all the respondents was to improve their living conditions. SSP water has helped to strengthen and also in some cases rekindle the life and economy of rural Gujarat.

Conclusions

The study draws the following major conclusions:

(a) Farmers in all the selected villages have benefited from the SSP water in terms of higher intensity of cropping pattern, better crops, lower costs of irrigation and predictable outputs. Most farmers are also willing to take crop loans. That water from SSP is resulting in higher production.

(b) Generally, the groundwater levels have increased and access to drinking water has become a lot easier. The study also points out to the implications of this for waterlogging and thus on productivity in the long run.

(c) The water has increased work wage opportunity and improved income in the selected regions.

(d) As far as gender issues are concerned, household income levels have gone up and access to drinking water has improved. At the same time new issues have emerged such as more milk animals and more burden of work for women.

Finally, one must find how to further improve the living conditions of people and build upon the positive changes brought about by the SSP waters. For this we can look to our past and observe what socio-economic or environmental factors enabled basic survival or led to complete disintegration in the long years of drought-like conditions.

Annex

Table A.1: Price of Different Crops Considered for the Assessment of Income (in Rupees)

S. No.	Name of the crop	Rate Per Quintal in Rs.
1.	Paddy	1000
2.	Wheat	1200
3.	Tobacco	800 to 1100 (depends on the quality)
4.	Cotton	2000 to 2200 (depends on the quality)
5.	Bajra	800
6.	Maize	800
7.	Tuver	2500
8.	Sesame	4500

24

Revisiting Sardar Sarovar Canal Based Drinking Water Project in Gujarat

Subhrangsu Goswami

Introduction

The Sardar Sarovar Dam Project on the Narmada river is one of the most debated projects, and so is the Sardar Sarovar Canal based drinking water supply project in Gujarat. The allocation of water for Gujarat from Sardar Sarovar Project has two distinct components—water for irrigation and water for domestic use and industry. While the irrigation component is well researched and debated by planners and policy makers at different levels, the drinking water component has always been in the limelight for mainly political reasons. This is probably because the drinking water component was included later during the 9th Plan to provide dependable water supply to water-stressed regions of the state. Once completed the project will cover more than 55 per cent of the villages and more than 65 per cent of the urban centres of the state. However, the project has so far been faced a lot of criticism. Even the need for it has been questioned time and again by some researchers and activists, with the argument that local sources can sustain the local demand, and Gujarat does not require centralised, capital-intensive projects like this.

The primary objective of this chapter is, therefore, to examine the need for the Sardar Sarovar Canal based water supply project in the context of overall status of water supply in the state and the changing strategies of the state government to cope with the pressing problem of water supply. The chapter also discusses the issues and challenges in implementing the project and achieving its holistic objectives, in the context of its built-inweaknesses, which arise from its concept and design on the one hand and the socio-cultural and administrative capabilities in the state on the other.

Section 1 – Status of Drinking Water in the State: Retrospection

Probably the first acute shortage of water in Gujarat was noted during the droughts of 1960-61. The first survey of drinking water was conducted in 1963-64 to study the problem. The survey reported that about 1,043 villages had no dependable facility for drinking water and 3,219 villages had inadequate supply of drinking water, that is, about 15-17 per cent villages suffered from drinking water shortages.

The problem acquired significant dimensions gradually with the increased use of groundwater in irrigation on the one hand and the declining traditional systems of managing local water supply on the other hand. In the Fourth Plan, the concept of No Source Village (NSV) was introduced to identify problem villages with inadequate supply of water, and accordingly a village was an NSV if it did not have a reliable source of drinking water.[1] The policy of the government was to identify such villages and provide them with a source. With the depleting groundwater resources, accompanied by fluctuating rainfall, the sources identified frequently turned out to be temporary. As a result, villages "with source" many times became "no source villages" when the temporary source dried up. The estimates of NSVs, therefore, fluctuated widely over time, from 5,086 in 1980, 4,833 in 1985, 16,351 in 1987-88 to 186 in 1990 and 416 in 1992 (Hirway and Patel, 1993). This indicated the fluctuating nature of the problem.

---

1. A village is a no source village if it has any of the following characteristics: (1) No public well, (2) has a public well that dries up in summer requiring villagers to travel more than 1 km to fetch water, (3) a source of water supply more than 1 km away, (4) no possibility of a well, needed a tube well for drinking water, (5) there is a public well, but the supply is below 70 lpcd, (6) non-potable water supply (GWSSB 2000).

Availability and Supply

Nature has endowed Gujarat with limited fresh water. The total renewable fresh water available, including the annual runoff from within the state and that allocated from the neighbouring states, and all the natural recharge of groundwater, is 54,593 MCM. This gives a per capita renewable fresh water availability of 1137 m^3 per annum for the year 2001. Therefore, as per water stress index[2], the state can be called "water stressed". But the availability of water is heavily skewed towards south and central Gujarat, which has 69.5 per cent of the total renewable fresh water. At regional level per capita availability in other regions is much less, at 427 m^3 in North Gujarat, 734 m^3 in Saurashtra and 875 m^3 in Kachchh (IRMA, 2001).

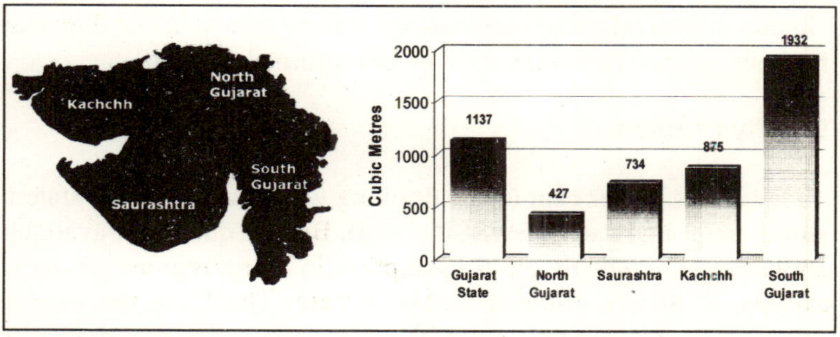

Map 1: Regionwise per Capita Water Availability Per Year
Source: IRMA, 2001.

In the past two decades, the state has lost about 27 per cent of its ground water resources, the loss being 50 per cent in North Gujarat. About 87 per cent area of the state has become "non-white" in groundwater, implying unsafe withdrawal of groundwater in these areas (Hirway, 1999). At the river basin level, the study done by

2. The Falkenmark water stress index measures per capita water availability and considers that a per capita water availability of between 1000 and 1600 m3 indicates water stress, 500-1000 m3 indicates chronic water scarcity, while a per capita water availability below 500 m3 indicates a country or region beyond the 'water barrier' of manageable capability (Falkenmark and Widstrand, 1992).

Amarasinghe *et al.* (2005), used disaggregated data to assess the water supply and demand across the river basins of India. According to the study Saurashtra, Kachchh and North Gujarat falls in the river basin named as westerly flowing river group 1 (WFR 1), which is the only basin of the country having physical water scarcity with a very high degree of groundwater development (132%) and abstraction ratio (194%), with depleted fractions of 92 per cent.

This water crisis was reflected in innumerable incidents of public rallies and demonstrations (sometimes resulting in violence) taken out to protest against water shortage in the different towns and cities as well as in villages of the state. The first major violent water riots were witnessed in Falla village of Jamnagar district (Saurashtra) in 1999. However, before that, in the 1990s, several demonstrations and rallies were witnessed in villages of Kachchh, Suarashtra and North Gujarat, in several urban centres in the state as well as in Gandhinagar, the state capital. Sometimes these rallies turned violent.[3]

Quality Problem

Today about 78 per cent of the drinking water supply in the state is based on groundwater (GWSSB, 2005). But the quality of available groundwater is not fit for consumption in many regions owing to excessive salinity, fluoride and/or nitrate. The State the longest coastline in the country, i.e. 1600 km and excessive withdrawal of groundwater from coastal aquifers has lead to ingress of seawater in the coastal aquifers rendering many thousands of drinking water and irrigation wells useless in coastal areas of Kachchh and Saurashtra (Bhatia, 1992; Kumar, 1995). Every year on an average 0.5 to 1.0 km distance from the coastline is affected by salinity ingress (Barot, 1996). As of February 2005 there were 2508 villages in the state affected by salinity in groundwater, which is not potable as per WHO standards (GWSSB, 2005). The latest data (Table 1) on the level of salinity in groundwater from coastal Saurashtra show that 379 villages out of a total of 954 (39.7% villages) are fully saline, i.e. the TDS content of groundwater is above 2000 mg/l throughout the year; 177 villages

3. As cited in Times of India (May 2, 2001; February 19, 2002; and January 18, 2003); *Indian Express* (April 6, 2002); *Economic Times* (April 9, 2002); *Gujarat Samachar* (May 4, 2002, May 18, 2003).

(18.6%) are partially saline, i.e. their groundwater becomes saline (>2000 mg/l) during the summer months. Thus one can infer that in about 60 per cent villages groundwater is not potable through the whole year.

Table 1: Level of Salinity in Groundwater in Saurashtra

District	*Coastal Villages*	*Number of villages*		*Percentage of Villages categorised as*	
		Partially Saline	*Fully Saline*	*Partially Saline*	*Fully Saline*
Amreli	92	23	22	25	23.9
Bhavnagar	198	35	50	17.7	25.3
Jamnagar	250	49	127	19.6	50.8
Junagadh	302	53	92	17.5	30.5
Porbandar	65	17	41	26.2	63.1
Rajkot	47	0	47	0.0	100.0
Total	954	177	379	18.6	39.7

Source: Salinity Ingress Prevention Circle, Rajkot (2005).

Many of the habitats in the state suffer from more than one water quality problem. It is evident from the recent data of Central Groundwater Board (CGWB), (Table 2), that out of 26 districts in the state, 20 are partly affected by salinity problem, 18 are partly affected by fluoride problem, 17 are partly affected by chloride problem, and 22 districts are partly affected by Nitrate.

In addition to the above problems, there are problems of water contamination arising from solid and liquid waste disposal from industries and human settlements. The regions around the major industrial centers like Vadodara, Bharuch, Ankleshwar, Vapi, Valsad, Surat, Navsari etc have polluted water sources, which have adversely affected their drinking water sources.

Impacts of Quality Problem

Salinity affected villages have high incidence of kidney stone compare to non-salinity affected villages. A study done by Rajnarayan Indu and Alka Rawal explains this between Mangrol and Maliya talukas of Junagadh district. The people of Mangrol taluka in its salinity-affected area are required to buy potable water for their daily use.

Table 2: Groundwater Quality Problems

Contaminants	*Districts affected (in part)*
Salinity (EC > 3000 µS/cm at 25 ° C)	Ahmedabad, Amreli, Anand, Bharuch, Bhavnagar, Banaskantha, Dahod, Porbandar, Jamnagar, Junagadh, Kachchh, Mehsana, Navsari, Patan, Panchmahal, Rajkot, Sabarkantha, Surendranagar, Surat, Vadodara.
Fluoride (>1.5 mg/l)	Ahmedabad, Amreli, Anand, Banaskantha, Bharuch, Bhavnagar, Dahod, Junagadh, Kachchh, Mehsana, Narmada, Panchmahal, Patan, Rajkot, Sabarkantha, Surat, Surendranagar, Vadodara.
Chloride (> 1000 mg/l)	Ahmedabad, Amreli, Bharuch, Bhavnagar, Banaskantha, Porbandar, Jamnagar, Junagadh, Kachchh, Dahod, Patan, Panchmahal, Sabarkantha, Surendranagar, Surat, Vadodara, Rajkot.
Nitrate (>45 mg/l)	Ahmedabad, Amreli, Anand, Banaskantha, Bharuch, Bhavnagar, Dahod, Jamnagar, Junagadh, Kachchh, Kheda, Mehsana, Narmada, Navsari, Panchmahal, Patan, Porbandar, Rajkot, Sabarkantha, Surat, Surendranagar, Vadodara.

Source: http://cgwb.gov.in/gw_profiles/st_Gujarat.htm\ (viewed on 20th October 2010).

The families affected by kidney stone have to incur high medical expenses and also have to lose wages during the time of the surgery to remove the stone. Besides all these there are other impacts on social aspects. There is typical resistance to marital alliances between the families of "salinity affected" villages and those not affected. This is not only due to poor water quality and fear of kidney stone but also to apprehensions about fetching drinking water from long distances, which the brides would like to avoid. This social impact is difficult to quantify (Indu and Rawal, 2007). On the other hand cases of villagers suffering from dental and skeletal fluorosis is quite high in the villages, where groundwater contains high levels of fluoride.

Section 2 – Coping Mechanism of State Government

After the first survey on the availability of drinking water in the state in 1963-64 the concept of No Source Village was introduced and the criteria for it were fixed. The policy was to find local sources of water supply for these villages. Since many state governments in the country were observed to be less than efficient in dealing with problem villages, the Government of India introduced the first Centrally Sponsored Programme, ARWSP (Accelerated Rural Water Supply Programme) in 1972-73 to cover problem villages under dependable sources of water supply. This 100 per cent centrally sponsored programme aimed at accelerating efforts of state governments in bringing to problem villages dependable sources of water supply. The ARWSP was followed by another centrally sponsored

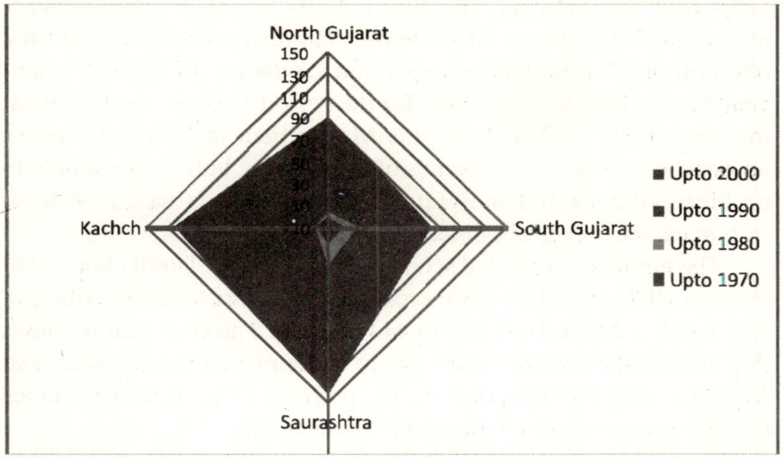

Chart 1: Regionwise commissioning of RWSS over the time

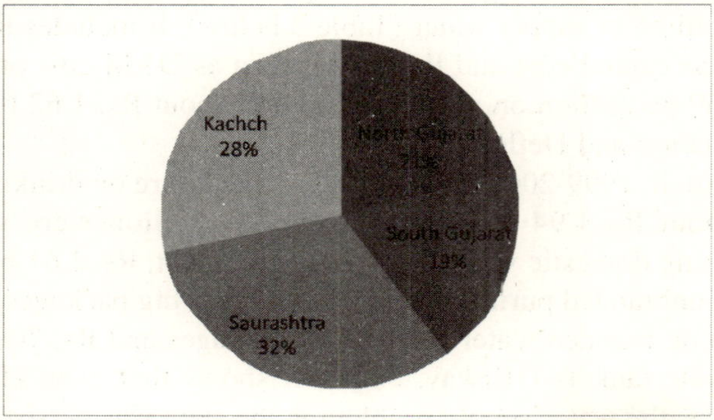

Chart 2: Regionwise Percentage Share of RWSS

Source: GWSSB, 2000

programme, the MNP (Minimum Needs Programme) in 1974-75 under which assuring potable and adequate water supply was identified as a minimum need to be fulfilled as a priority.

Role of Regional Water Supply Schemes

In the late 1970s and the early 1980s the state government introduced large regional water supply schemes to transfer bulk water supply to problem villages. Important among these were the Netherlands supported schemes and the World Bank assisted schemes were important among these. In the Sixth Plan the government spent Rs. 650 million on rural water supply and sanitation. In 1980-81 a ten-year Master Plan was designed for the period 1980-81 to 1990-91 at the cost of Rs. 5.72 billion. This Master Plan included two major approaches—Regional Schemes of transferring bulk water supply to problem villages and individual village schemes based on local resources, whenever possible.

The number of regional schemes was very small until about 1980 (nine in 1970 and 43 in 1980, each covering three to seven villages). It jumped to 209 in 1990 and to 444 in 2000. The chart below shows that most of the schemes (60%) were implemented in Saurashtra and Kachchh, which corresponds to the problem of local water sources in these regions in terms of quantity and quality.

As per the record of GWSSB (Table 3), during the last decade (1990-2000), the state government spent (1999-00 prices) about Rs. 31.54 billion to supply water (Table 3 below). It includes Rs. 25.41 billion as capital cost and Rs. 1.10 billion as O&M cost on RWSS, about 59.66 million on water tankers and about Rs. 1.62 billion on Desalination and Defluoridation technology.

Also, in 1999-2000 alone private expenditure on drinking water was about Rs 4.94 billion. Of this 2.14 billion were spent on purchasing domestic water purifying equipment, Rs. 1.61 billion on buying unbranded purified water in small and big packages, Rs. 880 million on branded water bottles and packages and Rs. 300 million on private tankers (Hirway, 2002). It shows that even after huge expenses, the water supply problem in the state was persistent.

Table 3: Increasing Expenditure of the State on Drinking Water Supply during 1990-91 to 1999-2000 at 1999-2000 Prices

(in Rs Crores)

Year	*RWSS*		*Water Tankers*	*RO/Desalinisation*		*Defluoridation*		*Grand Total*
	Capital	*O&M*		*Capital Cost*	*O&M Cost*	*Capital Cost*	*O & M Cost*	
1990-91	295.62	11.39	5.08	0	0	0	0	312.09
1991-92	201.18	13.72	22.6	0	0	0	0	237.50
1992-93	219.83	12.57	2.93	0	0	0	0	235.33
1993-94	157.5	12.04	13.07	0	0	0	0	182.61
1994-95	155.67	12.31	0	0.53	0	4.7	0.47	173.68
1995-96	158.05	11.22	1.01	0.41	0	1.91	0.19	172.79
1996-97	281.2.9	11.17	0.2	0.93	0.61	4.58	0.46	299.24
1997-98	288.2	8.7	0.13	0.88	0.57	19.34	1.93	319.75
1998-99	347.63	9.12	7.23	0.83	0.55	32.33	3.23	400.92
1999-2000	717.13	8.09	7.41	0	0	79.35	7.93	819.91
Total	2540.81	110.33	59.66	3.58	1.73	142.21	14.21	3,153.82

Source: GWSSB Reports, Master Plans of GWSSB for different years.

Experience of Rainwater Harvesting in the State

In the mid-1990s the state government set up a state level Recharge Committee to promote rainwater harvesting. In the Action Plan for the year 1996-97, the government decided to promote two rainwater harvesting schemes, namely, Roof Water Collection Tankas for households and the scheme of recharging groundwater through local rainwater harvesting structures at the community level. Rs. 10 million and Rs. 20 million were allotted to these two programmes respectively. So far more than 11,000 tanks and 500 groundwater recharge structures have been constructed in the state (GWSSB, 2003).

In 1998-99 the state government launched *Sardar Patel Sahbhagi Jal Sanchay Yojana* (SPPWCP) to promote the ongoing check-dam movement in the state, and particularly in Saurashtra, which is a highly suitable area for check-dams, with the rapidly flowing 70 rivers in the region. An independent evaluation of the Scheme (Shingi and Asopa, 2002) has examined, among other things, the impact of check-dams on the availability of drinking and domestic water supply in the supply in the surrounding regions of check dams. This evaluation study has observed, "an overwhelming majority of respondents indicated that check dams were able to reduce the severity of drinking water problem in their village." However, the study shows that this effect of check-dams does not last for more than one or at the most two drought years. One drought may change this situation dramatically.

Several NGOs in Gujarat also have experimented with rainwater harvesting methods to collect water to meet the local demand. Efforts have also been made to maximise recharge of groundwater on the land through a variety of water harvesting structures like nala plugging, percolation tanks, well recharge, link channels, percolation-*cum*-recharge wells, deepening of tanks and wells, reviving traditional water bodies etc., with the objective of maximising recharge of groundwater to increase the volume of sweet water available to people. Watershed development works, afforestation, etc., also fall in the same category of works as far as (ground) water recharge is concerned. In addition to the above, NGOs have also made efforts to create individual or community storages by constructing RRWHSs, sealing wells, constructing lined ponds, etc to isolate sweet water from saline water.

Is it Sufficient?

It is frequently claimed that rainwater-harvesting structures at the village level can collect enough water to take care of the local demand for water. However, with changing lifestyle and increasing demand, how far these water harvesting structures can help in resolving the problem of water supply in rural areas of Gujarat is a matter of concern.

- The demand for water per capita has increased today due to the changed lifestyle. With the population growth, the total quantity of water demand has increased multifold.
- Collecting water in traditional tanks, wells etc. have become difficult as the catchment area is frequently not available. Roads, buildings and other infrastructure have blocked the catchment areas.
- Groundwater aquifers are frequently empty, degraded or have water with excessive minerals or pollutants. Care will have to be taken while collecting rain water in safe aquifers.
- The demand for water for irrigation has increased considerably after the Green Revolution. Protecting water supply for people from irrigation (and other alternative uses) is a major task today. Farmers are always keen to use up collected water for irrigating their farms.
- In short, although harvesting rainwater is not a very difficult task, it is difficult to harvest it on an adequate scale, reserve some water for drinking, manage the quality of water and ensure its equitable distribution.

Scarcity Master Plan

At times the increased frequency and intensity of droughts, which were accompanied by shortages of drinking water, pushed the state to look for quick solutions to the problem. Preparation of scarcity master plan is one such approach adopted by the state. In the scarcity years many surface water sources in the state become dry. Even some of the small and medium dams start drying up from September. Slowly the larger dams also dry up in between December and January. Thus in almost every scarcity year the state government prepares a scarcity master plan. The major components of the crisis management are as follows:

- Feeding new and old pipelines (where sources have dried up) by drilling emergency bores and tube wells, whenever possible.
- Transferring water to problem areas by water tankers, water trains.[4]
- Drilling bores, tube wells etc. to access water from deeper aquifers for local population, whenever possible.

To obtain a clearer picture, the Scarcity Master Plan for the year 2004-05 is discussed below. It is observed that, during this period a total of 2,737 villages were declared water scarce in the state, of which 53 per cent were in Saurashtra and Kachchh. A sum of Rs. 402.4 million was spent with about 60 per cent share of Saurashtra and Kachchh. About 48 per cent of the total budget was spent for water supply through tankers alone. In case of Kachchh and Saurashtra, this share is as high as 65 per cent. These figures are self-explanatory of the fact that local sources during these scarcity periods in many areas proved to be absolutely defunct.

Experience of Technological Options in the State

The state government has launched several quality improvement programmes to improve quality of drinking water in problem areas. The major programmes are desalinisation and defluoridation. In the decade of the 1990s Rs 53.1 million were spent on desalinisation and Rs. 1.56 billion on defluoridation.

Desalination Technology

Gujarat was perhaps the first state in the country to take lead in installing desalination plants for brackish water conversion in rural areas. The first experiment in Reverse Osmosis (RO) Technology

4. In May 2000, to cope with the severe scarcity in the Jamnagar district, the state government decided to provide water through Water Train from Kodinar to Hapa about (400 km). About 600,000 gallons of water was supplied everyday from the month of May 2000 for two months. The cost of the water train was 80 million and the maintenance cost was about 7 million per month.

was done during the droughts of 1985-87. The R & D Programme of the GWSSB installed six RO plants during this period. Later, 11 more plants were added, of which one was installed by BARC and one by CSMCRI. Gujarat also received some plants under the Rajiv Gandhi National Drinking Water Mission.[5] However it was shocking that these plants did not work to expectation and could not supply potable water to the communities as expected. Probably desalination technology, being new for providing drinking water in rural areas, was less successful. Some of the factors responsible have been noted as improper site selection, improper plant design and problems pertaining to operation and maintenance (Shah *et al.*, 1997).

As far as large-scale plants are concerned no significant effort in this direction for domestic water supply has been made so far in the state. The experience of large-scale industrial desalination plants operating in Gujarat has been very encouraging. However, desalination technology in the state has always remained an emergency alternative.

It is also important to note that, while, the IPCC lists desalination plants as a supply-side adaptive measure available to meet potential increases in urban water demand associated with climate change (IPCC, 2001), it is evident that the desalinated water is expensive as compared to prevailing water supply mechanism practised in the state in terms of RWSS. It is also evident that the environmental footprint of desalination is significant. During the operating phase of the plants, one major environmental impact is the discharge of brine[6] on coastal or marine eco-systems, or in the case of inland brackish water desalination, on rivers, aquifers and soil. In short, decisions about exploring desalination in the state will revolve around complex evaluations of local circumstances and needs, economics, financing, environmental and social impacts, and available alternatives.

---

5. Government of India in Ministry of Rural Development launched a separate submission on "control of brackishness" under Rajiv Gandhi National Drinking Water Mission and total 240 Desalination plants were provided to various states for water supply in saline ones during 1990-94.
6. Brine is a concentrated salt solution that may be hot and may contain various chemicals.

Defluoridation Technology

The most widely used method of defluoridation is the method of removing excessive fluoride by the Nalgonda Technique usable at domestic and community levels. This method has been developed by NEERI, Nagpur. The state government of Gujarat installed 14 such community defluoridation plants in Amreli, Bhavnagar, Mehsana and Kachchh on an experimental basis. Later on, it installed 8 more plants under the Rajiv Gandhi Mission at the cost of Rs. 10.3 million. However, these plants are few and scattered, as it was not proved to be a long term for permanent solution. The state government appointed a committee in 1992 to examine the problem and to make suitable recommendations. The committee recommended the Nalgonda Technique as being usable at domestic and community levels for immediate relief. However, the committee also suggested setting up long distance pipelines to get potable water to problem areas. As a result, several regional pipeline schemes were designed to solve the problem of excessive fluoride in water.

Dew Harvesting

The prevailing view among the scientists in India (hydrologists, engineers, meteorologists) is that the dew is not of any utility to humans because of low quantity and infrequent occurrence. This view may not be as true for areas near coasts as for the hinterland (Sharan, 2007). Study done by Prof. Girja Sharan, at three locations along the Gujarat coast—Panandhro, Kothara and Mithapur—have shown that (a) dew occurs over a season of eight months—October to May, the quantity and frequency of occurrence are higher in summer than in winter months; (b) number of dew nights are large, varying from 95 to 105; and (c) the amount of dew water collected by specially made plastic condensers was equivalent of 20 mm, which in arid areas is appreciable. It is also argued that, dew is less in quantity than the rainfall but it is more uniformly distributed over the season which is twice as long as the rainy season in Kuchchh. In short, If suitable devices are deployed, as has been demonstrated by Prof. Girja Sharan, dew water can become a significant supplementary source for human use, in the coastal villages of Saurashtra and Kachchh.

Section 3 –Emergence of Narmada Water as a Long Term Solution

Looking to the severe depletion and degradation of water resources in the state, it was decided in the year 1990-91 (9th Plan) to use water from Narmada (i.e. Sardar Sarovar) to provide dependable water supply to the regions of Kachchh, Saurashtra and North Gujarat, covering 4,000 villages. The number increased to 8,215 villages and 135 towns in the late 1990s. In 2005 a revised Master plan for the project was prepared, which covers 9,633 villages and 131 towns from 17 districts. Some additions and deletions in number of villages have been made in this revised master plan. Many villages covered under the actual master plan was found to be uninhabited during project execution and were removed from the actual plan, where as some villages were added looking into the scarcity or demand from their habitations.

The main objectives of the project are as follows:

- To supply adequate and safe drinking water to the areas suffering from severe shortage of water supply.
- To feed the regional water supply schemes, which are not providing adequate water supply due to failure of their sources.
- To provide water supply to those regional schemes and individual (village) schemes which provide poor quality of water supply due to various reasons.
- To supply potable water to those towns and villages which have excess salinity, excess fluoride or excess nitrates in their water supply.
- To reduce out migration of people by ensuring drinking water and by promoting employment in the areas suffering from water shortages.

Project Coverage

The Sardar Sarovar Canal based Drinking Water Pipeline Project is perhaps the biggest drinking water project in the world. As per the revised master plan this project will cover the whole of Saurashtra and Kachchh as well as parts of North Gujarat (Ahmedabad, Mehsana, Banaskantha, Sabarkantha, and Panchmahal). The capital cost of the

Table 4: District wise Coverage of Villages/Towns in Revised Master Plan

Sl. No.	Districts	Number of Villages in the district	Villages covered in the Master Plan		Number of Towns in the district	Towns covered in the Master Plan	
			Number of villages	Percentage of Villages		Number of towns	Percentage of Towns
1.	Ahmedabad	546	546	100.0	25	25	100.0
2.	Amreli	615	615	100.0	8	8	100.0
3.	Anand	350	0	0.0	12	0	0.0
4.	Banaskantha	1,244	816	65.6	6	2	33.3
5.	Bharuch	657	0	0.0	8	0	0.0
6.	Bhavnagar	790	790	100.0	13	13	100.0
7.	Dahod	693	0	0.0	4	0	0.0
8.	Dangs	311	0	0.0	0	0	0.0
9.	Gandhinagar	291	291	100.0	10	4	40.0
10.	Jamnagar	698	698	100.0	16	16	100.0
11.	Junagadh	923	923	100.0	12	12	100.0
12.	Kachchh	886	886	100.0	8	8	100.0
13.	Kheda	612	65	10.6	8	0	0.0
14.	Mehsana	593	464	78.2	8	7	87.5
15.	Narmada	552	78	14.1	3	0	0.0
16.	Navsari	374	0	0.0	9	0	0.0
17.	Panchmahal	1,201	408	34.0	6	3	50.0
18.	Patan	517	421	81.4	5	5	100.0
19.	Porbander	182	182	100.0	5	5	100.0

Sl. No.	Districts	Number of Villages in the district	Villages covered in the Master Plan		Number of Towns in the district	Towns covered in the Master Plan	
			Number of villages	Percentage of Villages		Number of towns	Percentage of Towns
20.	Rajkot	844	844	100.0	10	10	100.0
21.	Sabarkantha	1,372	956	69.7	8	6	75.0
22.	Surat	722	0	0.0	15	0	0.0
23.	Tapi	445	0	0.0	3	0	0.0
24.	Surendranagar	650	650	100.0	7	7	100.0
25.	Vadodara	1,548	0	0.0	16	0	0.0
26.	Valsad	450	0	0.0	17	0	0.0
	Total	18,066	9,633	53.3	242	131	54.1

project has been estimated to be Rs. 103.7 billion (2010 prices), which includes investments in trunk mains, water treatment plants, pumping machinery, civil works and water distribution. The project includes 48 bulk schemes, 123 distribution groups and a 3,250 km long pipeline network.

The project is extremely important for Gujarat, as it will cover 53.3 per cent villages and 54 per cent urban centers of the state. It will cover 100 per cent villages in Ahmedabad, Amreli, Bhavnagar, Gandhinagar, Jamnagar, Junagadh, Kachchh, Porbandar and Surendranagar districts. Similarly all towns are proposed to be covered in Ahmedabad, Amreli, Bhavnagar, Jamnagar, Junagadh, Kachchh, Patan, Porbandar, Rajkot and Surendranagar.

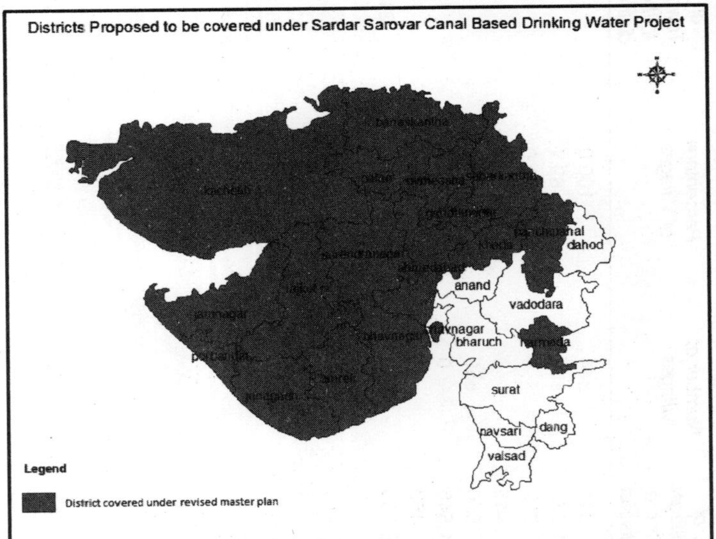

Map 2: Districts to be Covered under Revised Master Plan of Sardar Sarovar Canal based Drinking Water Project

In this project, water will be transported through bulk water transmission pipelines and distribution network to the doorsteps of villages and towns, which do not include in-village/town drinking water supply infrastructure development. Thereafter, in rural areas Panchayats will take over the responsibility of distribution, management, operation and maintenance of in-village water supply systems whereas local urban bodies will be responsible for distribution management, operation and maintenance in towns.

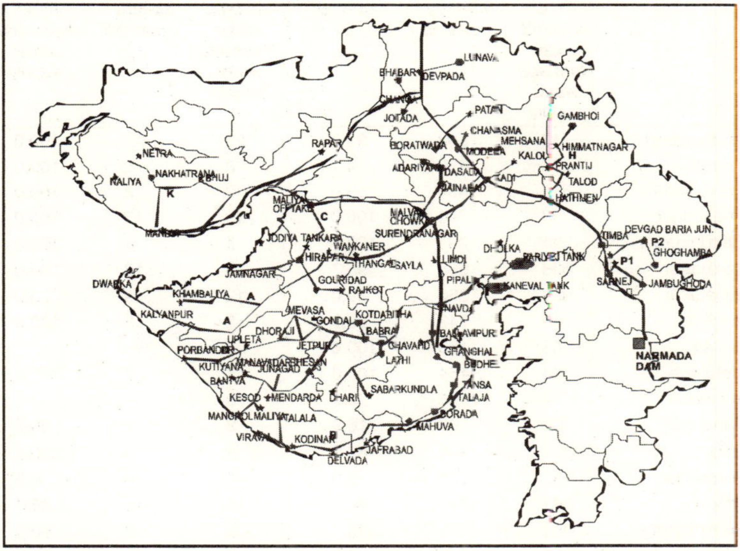

Map 3: Sardar Sarovar Canal based Drinking Water Pipeline Layout Plan

Source: GWSSB, 2001.

The whole project is handled mainly by three organisations: Gujarat Water Supply and Sewerage Board (GWSSB) is responsible for implementing the Distribution Network, Gujarat Water Infrastructure Limited (GWIL) is responsible for bulk transmission of water from Narmada to the villages and towns through various pipelines, and Water and Sanitation Management Organisation (WASMO) is responsible for promoting water committees (Pani Samitis) for local management of water supply.

Progress So Far

Though during its initial phase the project was quite slow in terms of implementation, by the end of October 2010 it has covered 6,513

Table 5: Actual Coverage of Towns and Villages till October 2010

District	*Number of villages to be covered under Narmada water supply*	*Number of Villages already covered*	*Percentage of Villages already covered*	*Number of Towns to be covered under Narmada water supply*	*Number of Towns already covered*	*Percentage of towns covered under Narmada water supply*
Ahmedabad	546	508	93.0	25	25	100.0
Amreli	615	456	74.1	8	8	100.0
Banaskantha	816	454	55.6	2	2	100.0
Bhavnager	790	790	100.0	13	13	100.0
Gandhinager	291	291	100.0	4	4	100.0
Jamnager	698	668	95.7	16	16	100.0
Junagadh	923	360	39.0	12	9	75.0
Kachchh	886	743	83.9	8	8	100.0
Kheda	65	65	100.0	0	—	—
Mehsana	464	464	100.0	7	7	100.0
Narmada	78	31	39.7	0	—	—
Panchmal	408	69	16.9	3	2	66.7
Patan	421	348	82.7	5	5	100.0
Porbander	182	11	6.0	5	0	0.0
Rajkot	844	541	64.1	10	6	60.0
Saberkantha	956	271	28.3	6	3	50.0
Surendranager	650	443	68.2	7	6	85.7
Total	9633	6513	67.6	131	114	87.0

Source: GWSSB, October 2010.

villages, i.e. about 68 per cent villages against proposed coverage, and 114 towns i.e. about 87 per cent against the proposed coverage. This is a commendable achievement. A 2,170 km pipeline has been laid (which is about 67% of the proposed 3,250 km pipeline). Quarterly progress in terms of coverage of towns and villages under the project during 2009-10 is shown in the graph below (Graph 1).

The graph 2 shows the districtwise coverage achieved till October 2010. It is evident that 100 per cent towns are already covered in Ahmedabad, Amreli, Banaskantha, Bhavnagar, Gandhinagar, Jamnagar, Kachchh, Mehsana and Patan district. On the other hand 100 per cent villages have been covered in Bhavnagar, Gandhinagar, Kheda and Mehsana.

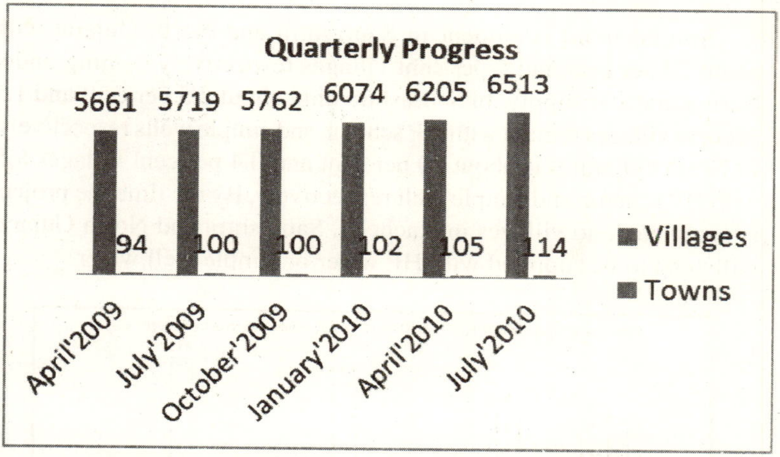

Graph 1: Quarterly Progress in Coverage of Settlements

Source: GWSSB, October 2010.

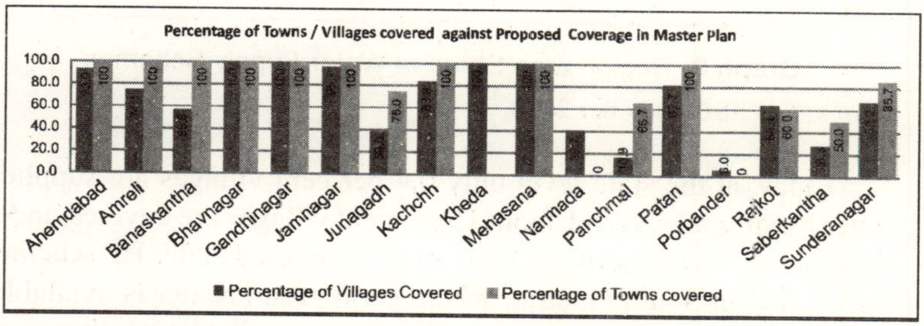

Graph 2: Actual Coverage against Proposed Coverage

Source: GWSSB, October 2010.

Changing Face of Rural Water Supply in the State

It should be noted that in Kachchh, Saurashtra and North Gujarat, the majority of villages are now covered by Narmada water (Table 6). In Kachchh about 84 per cent villages are supplied by Narmada water and only about 5.5 per cent villages are covered by regional water supply schemes (RWSS) and about 10.6 per cent villages are covered by individual piped water supply schemes. In short in Kachchh all villages are supplied with piped water and not a single village is supplied by a HP scheme or simple well.

Similar trend is evident in Saurashtra and North Gujarat with about 70 per cent and 5 per cent villages respectively coming under Narmada water supply. In Saurashtra only about 2.5 per cent and 1.5 per cent villages remain with HP scheme and simple wells respectively. In North Gujarat it is about 10 per cent and 0.4 per cent villages left with HP scheme and simple well respectively. By the time the project is completed, no villages in Kachchh, Saurashtra and North Gujarat will need to be supplied with HP water or simple well water.

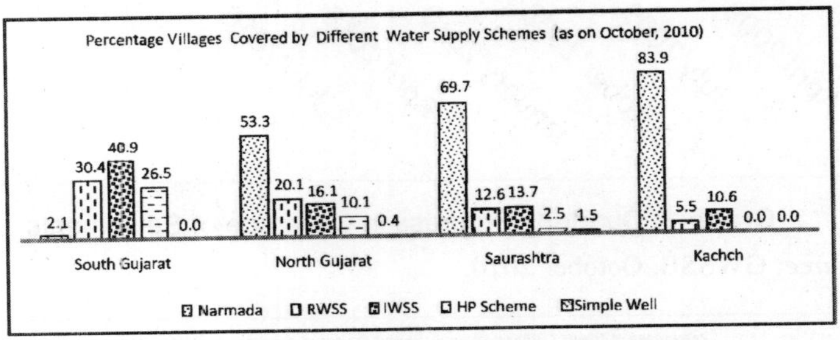

Graph 3: Region wise Share of Water Supply Schemes

Source: GWSSB, October 2010.

Today, at the state level only 0.4 per cent villages are supplied water by simple well and about 15 per cent villages are covered under HP scheme. The majority of such villages covered under HP schemes are located in South Gujarat, where good groundwater is available. In short, good quality water supply is now available in villages of Kachchh, Saurashtra and North Gujarat thanks to Sardar Sarovar Canal based drinking water project.

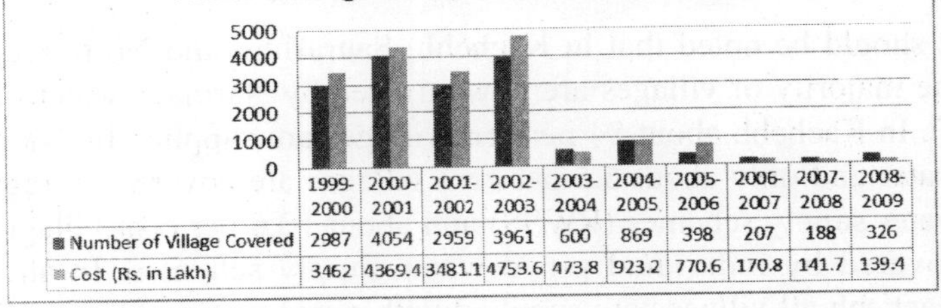

	1999-2000	2000-2001	2001-2002	2002-2003	2003-2004	2004-2005	2005-2006	2006-2007	2007-2008	2008-2009
Number of Village Covered	2987	4054	2959	3961	600	869	398	207	188	326
Cost (Rs. in Lakh)	3462	4369.4	3481.1	4753.6	473.8	923.2	770.6	170.8	141.7	139.4

Graph 4: Changing Coverage and Expenditure on Tanker Water Supply in the State

Source: GWSSB, 2009.

Table 6: Coverage of Narmada Water and Other Water Supply Schemes

Districts	*Total Number of Village in District*	*Villages Covered as on October 2010*				
		Narmada Canal Based Water Supply Projects	*Regional Water Supply Schemes*	*Individual Piped WSS*	*HP Schemes*	*Simple Well*
South Gujarat	7915 (100)	165 (2.1)	2408 (30.4)	3241 (40.9)	2101 (26.5)	0 (0.0)
North Gujarat	5213 (100)	2779 (53.3)	1049 (20.1)	841 (16.1)	525 (10.1)	19 (0.4)
Saurashtra	4052 (100)	2826 (69.7)	509 (12.6)	556 (13.7)	100 (2.5)	61 (1.5)
Kachchh	886 (100)	743 (83.9)	49 (5.5)	94 (10.6)	0 (0.0)	0 (0.0)
Gujarat Total	18066 (100)	6513 (36.1)	4015 (22.2)	4732 (26.2)	2726 (15.1)	80 (0.4)

Note: Figures in bracket are in percentage

Source: http://www.gwssb.org/pdf/narmadaprojects.pdf, viewed on 20th December 2010.

Data for the last ten years of villages covered by tanker water supply in the state and the cost of the supply are presented in the graph below. It is evident from the reduced number of villages covered that the expenditure attached to that has decreased substantially in recent years. During 2002-03, a total of 3,961 villages were covered under tanker water supply at an expense of about Rs. 475 million, whereas in 2008-09 only 326 villages were supplied by tanker spending only about Rs. 14 million.

Changing Face of Urban Water Supply in the State

Urban water supply problems differ from the rural in several ways. A major difference is the per capita requirement of water supply, which is much higher in urban areas, ranging from 135 lpcd to 250 lpcd, based on the class of city and the availability of drainage facility.

While the absence of perennial rivers is a matter of continuing worry for Gujarat, the ever increasing population and demand for more

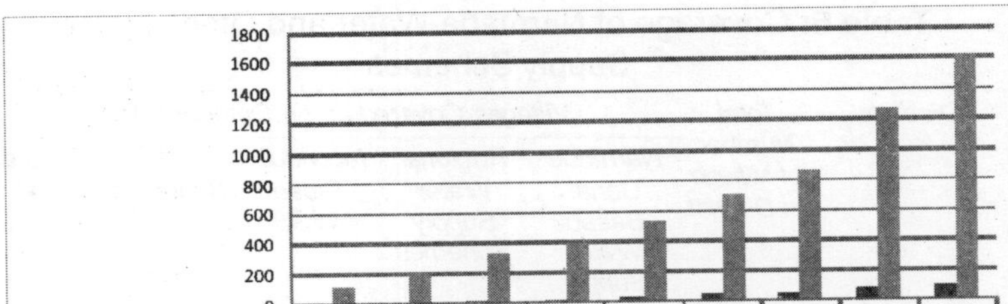

	1951	1961	1971	1981	1991	2001	2011	2021	2031
Population (in lakh)	8.37	11.5	15.86	23.82	33	44.27	53.99	79.58	101.44
Water Demand(in MLD)	99	209.1	331.6	397.5	528	708.3	863.8	1273.3	1623

Graph 5: Increasing Population and Water Demand in the City

Note: Figures of 2011, 2021,2031 are projected
Source: AMC, 2010.

water supply in urban areas have pushed urban local bodies to exploit groundwater sources. However these are not adequate to cater to drinking water demand or overcome the drinking water supply problem. Overexploitation of groundwater in urban areas in the state over the time has resulted in failure of such sources in quality as well as quantity.[7]

Water supply in about 77 per cent urban centres including 28.6 per cent towns that use bulk water from distant sources, is from groundwater. It is estimated that about 87 per cent of urban water supply in the state comes directly or indirectly from groundwater (Hirway & Lodhia, 2004). Therefore, bulk transfer of water from distant places has emerged as an important means source of water supply in urban areas. While the Sardar Sarovar canal based drinking water project aims to cover 131 urban centres in the state, it is important to see how Narmada water can change the face of urban water supply in the state. To examine this aspect the case of Ahmedabad city is discussed in the following section.

Case of Ahmedabad City

Ahmedabad is the seventh largest metropolis in India and the largest city of Gujarat. The population of this city has increased from 3.5

7. http://www.gwssb.org/pdf/narmadaprojects.pdf, viewed on 20th December 2010.

million in 2001, within city limits covering 190 sq. km, to the currently estimated 5.6 million in limits of 466 sq. km. Graph 5 shows the increase in population and demand for water in the city since 1951.

Since the year 1891, the main source of water supply for Ahmedabad city was the non-perennial Sabarmati river. In response to the growth of the city area and water demand, the Dudheshwar water works on the river bank were constructed in 1950 to supply 68.0 MLD water. With increase in population and water demand, the Dudheshwar facility was upgraded by adding capacity of 45.4 MLD with six jack wells in the river bed. Thereafter, 30 infiltration wells were added in the river bed to supply an additional 68.10 MLD of water during the years 1950 to 1970.

During this period, a number of bores/tube wells were also sunk to fulfil the ever-increasing water demand of the city. Thus, during the period 1970-1980, groundwater accounted for 87 per cent of the city's total water supply, surface water from the river was also diminishing owing to the failure of rains in its catchment area.

Considering the future water demand of the city, the Dharoi reservoir-based water works at Kotarpur was planned. However, it was subsequently observed that over a period of 21 years, there were only eight years when the Dharoi reservoir got filled to capacity. Due to insufficient and erratic rainfall pattern in the catchment area, it was not fully replenished in the remaining years.

As an interim solution, six French Wells were constructed in the Sabarmati river bed to draw 182 MLD of sub-soil water during the period 1985-2000. However, the water from these French Wells and Dudheshwar water works was inadequate to fulfil the increasing water demand of the growing city. Thus it had to depend heavily on groundwater.

Status of Groundwater in the City: Of the 11 talukas of Ahmedabad District, Ahmedabad City and Daskroi have continued to be the over-exploited talukas. They have the highest level of groundwater development, which has increased from 148.63 per cent in 2002 to 150.86 per cent in 2005. Continuous pumping of groundwater by AMC as well as numerous private tubewells has resulted in rapid depletion of the water table to about 100 m in the year 1975. Receding at 3 to 4 metres per year, the table touched 300 metres in the year 2000.

As per data available from CGWB, the natural recharge in this area is only 13 per cent of the total precipitation received. It is evident that there is almost no impact on the groundwater levels in the city even after the monsoon because of the low rate of natural recharge Graph 6 shows the average groundwater levels in the AMC tubewells and their variation in last 45 years.

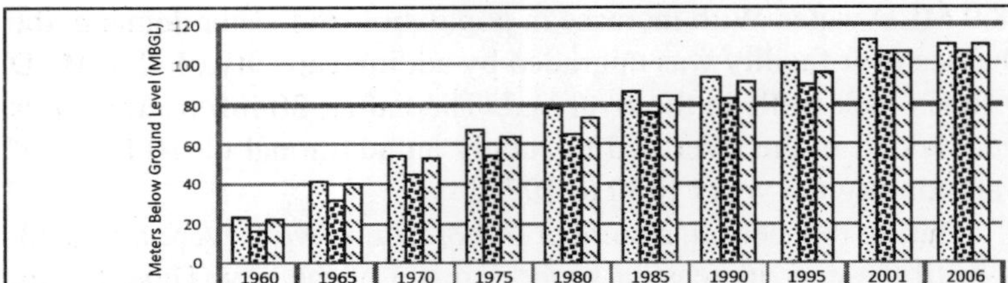

	1960	1965	1970	1975	1980	1985	1990	1995	2001	2006
Central Ahmedabad	24.4	41.5	54.9	67.1	78	86.6	93.9	100	112.8	109.7
East Ahmedabad	17.1	31.7	45.1	54.9	65.8	75.6	82.9	90.2	106.7	106.7
West Ahmedabad	23.2	40.2	53.6	64.6	73.8	83.5	91.4	96.3	106.7	109.7

Graph 6: Pumping Water Levels in AMC Tubewells (in mbgl)
Source: AMC, 2006.

The increasing dependency on groundwater in the city has a twofold direct impact. With decreasing groundwater level, the cost of pumping it is increasing. The frequent deepening of tubewells and borewells is another component of incremental expenditure. On the other hand due to inadequate surface water many citizens are forced to consume groundwater with TDS content as high as 2500 mg/l. Many studies have also recorded large expenditure on the purchase of water purifying equipment at the household level.

Impact of Water Transfer: In April 2000, AMC brought additional 300 MLD of water from Shedhi branch of Mahi canal through Raska weir, by spending Rs. 1.20 billion. Finally, the Narmada Canal gravity pipeline of 2,500 mm dia. MS pipe from Narmada canal to Kotarpur water works for drawing 330 MLD of water was also conceptualised, completed and commissioned in January 2007 under the JNNURM Programme. To cater the drinking water to the new western part of AMC, the Dholka branch of Narmada canal was also tapped to draw 400 MLD of water and 275 MLD capacity of water treatment plant at Jaspur which was constructed and commissioned in the year 2006-07.

It is evident that water levels in the city are now slowly recovering from the previous decline since 2003 onwards. This phenomenon of recharging of the groundwater leading to a rise in water levels can be attributed to the availability of Narmada Main Canal Water in Sabarmati River (Joshi, 2006).

As can be seen in the graph below, During the period 1970-1980 the consumption of groundwater in the city reached 87 per cent of its total water supply. This was because the surface water from the river was rapidly diminishing. However, with concrete steps taken by AMC to harness surface water sources (Mahi and Narmada Water), the utilisation of groundwater has been substantially reduced. Over the last 5 years dependency on groundwater has drastically reduced with only 5-10 per cent groundwater utilised. This has directly impacted the health of citizens in terms of decreasing water borne diseases in the city.

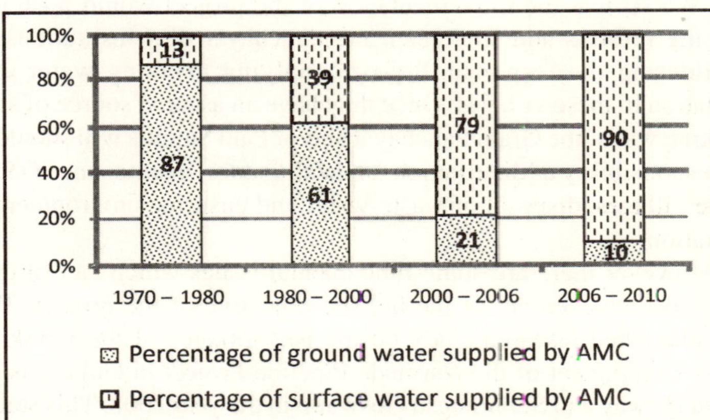

Graph 7: Decreasing Share of Groundwater Supply in the City
Source: AMC, 2010.

It is expected that by 2010-11 AMC will completely stop the use of groundwater. The multifold benefits of this transformation are already visible:

- The existing groundwater quality indicates presence of TDS of 1300 mg/l. Hence, by stopping the groundwater resources, the population will get water with TDS of about only 180 mg/l. This will prevent various diseases related to digestive system, kidneystone, heart problems, blood pressures etc.

- Electrical energy will be saved by stopping the pumping of groundwater.
- Groundwater table in this area will rise and hence the quality of groundwater will improve. Thus, groundwater sources can be utilised in case of emergencies, water scarcity, breakdown in present supply etc. (AMC, 2010).

Challenges in Achieving Expected Benefits

This project is perhaps the world's largest pipeline water supply project with the claim to features like financial sustainability, an improved operational system and institutional sustainability. It is also claimed by GWSSB that the implementation of the project would push the ongoing reforms and strengthen the Panchayati Raj Institutions in shouldering the responsibility of supplying drinking water and sanitation in these villages. Once they have an assured source of safe drinking water, the Gram Panchayat and/or Pani Samitis will shoulder the responsibility of distribution, collecting dues, maintenance O&M in the villages, disposal of waste-water and ensuring environmental sanitation.

However there are some fundamental issues which can further delay the achievement of the holistic objective of the project. The first published research study on the performance of the Drinking Water Component of the Narmada Pipeline Project in Gujarat is by Indira Hirway and Subhrangsu Goswami in the year 2008. This study, based on the initiative of participatory monitoring system to assess implementation of the Narmada Drinking Water Pipeline Project, claims that, despite some positive features, the project does not seem to be doing very well. It has not been able to ensure adequate water supply in a regular and dependable manner in all the regions covered (Hirway & Goswami, 2008). Over time many NGOs have also raised similar issues and concerns. Some such issues are discussed in the section below, which probably need to be addressed at the earliest.

The Sardar Sarovar Canal based water supply project primarily follows the principle of long distance water transfer and distribution, followed in many regional water supply schemes in the state. It is, therefore, important to examine the strengths and weaknesses of such schemes. Several studies have examined their working—Usha Sharma's study of randomly selected 29 RWSSs located all over the state (Sharma, 1995), the Haskoning study of the Netherlands supported RWSSs in Banskantha, Mehsana, Amreli and Bhavnagar (Haskoning, 1999), the CEPT study of Lathi-Liliya RWSS in Amreli, the DANIDA study of two RWSSs in Banaksantha and Mehsana (DANIDA, 1996), and the CAG Report (2001).

A major finding of all the studies is that the village level availability of water supply is not satisfactory. They show that (1) tail-end villages are usually deprived of water supply, (2) for the other villages also the water supply is often irregular and unreliable, (3) the quantity of water supply is often inadequate (less than 10 lpcd sometimes), and (4) the quality of water is not potable either because of problems at the source or because of contamination due to leakages and breakages. Some similar issues observed in Narmada water supply are discussed below.

Unreliability of Supply

According to the guidelines all villages covered under the Narmada project are expected to receive water every day. However, regularity of supply is questioned in many villages where water sometimes reaches once in two or three days. While GWSSB officials blame the unexpected power-cuts, failure of pumps and breakage of pipelines as major reasons for this, they should have alternative plans to take care of such problems to strengthen the reliability of supply.

Inadequacy of Supply

According to the norms, each village is expected to get water at the rate of 70 lpcd. In the absence of installed water meters, it is difficult to assess the adequacy of water supply. Generally the technicians (valve men) try to estimate it based on the duration of supply. Based on population at village level it varies from one hour per day to 3 hours per day. With leakages in the pipeline resulting in low pressure, sometimes a 3 hour long supply effectively becomes equivalent to normal supply of one hour to the end users. Therefore, it is essential to install a water meter at every village to assure quantum of supply.

Unequal Inter-Village Distribution

It is in general observed that, It has not been possible for the authorities to provide equitable water supply to all villages, because of frequent breakages and leakages in the inter-village pipelines due to technical problems such as weak joints, uneven pressure and design problems; poor maintenance of pipelines; failure of pumping stations or of main head works; electricity-failure; and deliberate breakages and theft of water by farmers, cowherds and others.

Unequal Intra-Village Distribution

As already discussed, in this project, water will be provided at the doorsteps of villages and towns. It does not include in-village/town drinking water supply infrastructure development. While in urban areas many local bodies are able to manage distribution, in many villages it is observed that distribution to the *falias* within them is not equitable. In several villages, intra-village distribution of water is completely dominated by people from so called upper castes. This social barrier cannot be overlooked, when some villages are deprived of any other source of water supply.

Inefficiency in Operation and Maintenance

The O&M work has been entrusted to private contractors who appoint linesmen to do the job. The duties of a linesman are to operate the valve, maintain the pipeline and clean the tank periodically, report breakages/leakages and manage its repair, observe the quality of water and apply chlorine if required, and support the pani samiti/village panchayat in managing the Narmada water supply. However, it is observed that contractors are always keen to reduce their costs by understaffing (Hirway & Goswami, 2008). Adequate supervision of the contractors by the GWSSB officials is needed.

Water Charges, Recovery and Financial Viability

While water charges under this project are meant to cover the O&M costs and to provide for meeting the capital expenses for future replacement of the system, the status of recovery in the state has so far not been very encouraging. According to GWSSB officials, current recovery varies between 25 to 35 per cent. The status of recovery should also be viewed against the fact that the cost of Narmada water per unit is in fact much higher than what is charged to villagers. Therefore, even if cross, subsidisation is done to reduce the price, the present charges will fall tremendously short of the costs.

Neglected Local Sources of Water Supply

It is evident that the number of villages in the state using different type of local sources, is not insignificant. However, in these villages, after Narmada water having become available, local sources are being badly neglected. Similar past experience has been shared by Usha Sharma in her study, which concludes that, the present RWSS approach positively discourages the survival and growth of local water sources (Sharma, 1995). Even villages suffering from serious shortages have neglected their own traditional tanks and wells. This neglect of local water resources must be discouraged. There is definitely a need to encourage their conservation and augmentation even if they are considered inadequate to meet the total local demand.

Absence of Systematic Disposal of Used Water

Any amount of domestic water consumption ends with generation of waste water. In normal conditions for domestic water supply it is observed that about 80 per cent of the used water is discharged as waste water. Therefore any village with a population of 2,000 with 70 lpcd water supply will generate waste water amounting to 112,000 lt. per day. However, management of waste water at the village level seems to be not so efficient. It is a well established fact that proper disposal of waste water can reduce the incidence of waterborne diseases. The project should encourage better management of waste water at the village level in terms of its collection and systematic disposal with required treatment.

Inefficient Village Level Water Management

The state government, through its circular of 24 December, 2002, has made the formation of Pani Samitis mandatory for all village

Panchayats in the state. The Panchayat is supposed to form the Samiti by passing a resolution. Each Pani Samiti will have 10-12 members, of whom one third will be women, 5 from the village Panchayat, 2-3 from locally registered NGOs (like local cooperatives, Mahila Mandal, Farmers' society etc.,) and 1-2 from SC/ST inhabitants.

While the Pani Samiti can play a vital role in the village level distribution and management of water and sanitation, it is observed that many villages have not yet been able to form them. There are villages where Pani Samitis exist only in name. Therefore, the highest priority should be given and suitable instrument devised to make Pani Samitis work efficiently. Once this is done, Pani Samitis will be capable of shouldering responsibilities like (1) taking care of water and sanitation management in the village, (2) identifying local water sources and preparing a scheme for their development for implementation, (3) managing the water sources of the village, (4) carrying out O&M of the regional scheme if local sources are not available, (5) taking care of sanitation and cleanliness of the village, (6) assuring equal distribution of water in the village, and (7) fixing and collecting water charges from people.

Summing Up

It is evident from the study that Sardar Sarovar Canal based drinking water project in Gujarat is of immense importance. The state has been suffering from chronic problem of drinking water supply since 1960s. Over time, with increasing demand for drinking water, the concerns become more critical. The problem has multiple facets in terms of limited availability of fresh water resources, quality problem, sectoral competition as well as regional dimension.

With full understanding of the changing nature and scale of the problem, the approach of the state government has also changed substantially. To start with the state government and NGOs have experimented sufficiently in rain water harvesting (direct harvesting as well as recharge of groundwater), to meet the local demand. However, with the climate variability, increasing demand, changing lifestyle, and sectoral competition, in many instances these efforts have proved to be unsustainable in the long run.

The state government has launched several quality improvement programmes for drinking water in the problem areas. Installation of Desalination and Defluoridation technology are two major technological interventions. However, these technologies in the state have always remained an emergency alternative.

Periodically the state government has prepared scarcity master plans which always focus on quick solutions like drilling emergency bores and tubewells, transferring water to problem areas by water tankers, water trains, sinking bores, tubewells etc. However, these quick solutions often proved unsustainable in long term. Therefore, the state government, over time has started relying more and more on RWSSs. However with limited fresh water resources and increasing demand for it over time, the existing RWSS schemes in the state are also proving inadequate. Therefore, although the SSP did not have initially include a component for drinking/domestic water supply, considering the severe depletion and degradation of water resources in the state, it was added to provide dependable drinking water supply to the regions of Kachchh, Saurashtra and North Gujarat.

Sometimes the need for the SSP has been questioned by groups of researchers and activists. The project has also been criticised with concerns about unequal distribution and poor management. During its initial phase the project was quite slow in terms of implementation, however by the end of October 2010 it has covered 6,513 villages (68%), and 114 towns (87%) against the proposed coverage, which is a commendable achievement. As far as rural areas are concerned, today in Kachchh 100 per cent villages are covered with piped surface water supply. In Saurashtra and North Gujarat only about 4 per cent and 10 per cent villages respectively are supplied from groundwater sources. This status will improve shortly with 100 per cent coverage of proposed villages under the master plan. In urban areas, the impact in terms of increased per capita supply and reduced dependency on groundwater is evident in many cities.

While this project is perhaps the world's largest pipeline water supply project claiming features like financial sustainability and improved operational system and institutional sustainability, this chapter observes that, like any other long distance pipeline scheme under which water is transported through pipelines, this project also has several built-in-weaknesses, which arise from the project concept and design on the one hand and the socio-cultural and administrative capabilities in the state on the other.

These weaknesses are of varying nature, and raise questions on the reliability and adequacy of supply, equitable inter-village and intra-village distribution, efficiency in Operation and Maintenance, water charges, recovery and financial viability, sustainability of local sources of water supply, and above all the capacity and efficiency of village level water management. Overcoming these weaknesses should be a major concern of policy makers and implementers at present, to achieve the long term sustainability of the project.

25

Urban Drinking Water Security and Sustainability in Gujarat

Meera Mehta and *Dinesh Mehta*

Introduction

With rapid economic development, Gujarat has also experienced rapid urbanisation. The share of urban population has grown from 22 per cent in 1901 to 37 per cent in 2001, making it the third most urbanised state in the country. Trends over the last decade suggest that in 2011 the urban population is likely to be nearly 50 per cent of the state's total population in 2011 (Dave, 2010).[1] The recent McKinsey report (MGI, 2010) on India's urbanisation states that Gujarat will be 66 per cent urban with 48 million urban residents in 2030.

One of the key challenges of urbanisation in Gujarat is provision of drinking water. A large part of the state is water stressed and has severe shortage of drinking water. In this paper, we attempt to show that with Narmada water reaching the interiors of Kachchh, North Gujarat and Saurashtra, and the Government of Gujarat's priority to this sector, there is now a possibility of drinking water security for urban areas of the state. Narmada has obviated the need to transport water by tankers and trains during the harsh summer months.

---

1. An informal assessment from the initial house listing exercise for the 2011 Census suggests that urban population in the state may already be close to 50 per cent in 2010. This would mean doubling of urban population in Gujarat from 18 million to 36 million in a decade. *Times of India*, Ahmedabad, September 23, 2010.

However, with rapid growth of urban population and rising consumption of water due to change of life style in urban areas, it is likely that the drinking water stress will increase. The threat to water security is also partly due to inefficient water management and governance. In this chapter, we focus on two major issues that affect sustainability of water—non-revenue water and cost recovery. For both these issues, we describe the current situation and discuss the potential ways forward to increase sustainability of urban water supply services. In describing the current situation, we use information from the studies under the Performance Assessment System (PAS) Project.[2]

Increasing Urbanisation in Gujarat

Gujarat's urban population of the state has been rising steadily since the formation of the state in 1960. The pace of urbanisation, as measured by the decadal growth of urban population, was significantly higher than the national average during the decade 1991-2001, and is expected to remain high in the next two decades as well.

In 2001, over 50 per cent of the urban population in Gujarat resided in the seven cities that are classified as municipal corporations. This concentration of urban population increases to nearly 60 per cent if urban agglomerations around these cities are taken into account. Nearly 40 per cent of urban population of Gujarat resides in the three large cities of Ahmedabad, Surat and Vadodara.

Table 1: Urbanisation Levels and Growth Trends

Year	Gujarat			India		
	Urban Population (Million)	Share of Urban population (%)	Decadal Growth Rates (%)	Urban Population (Million)	Share of Urban Population (%)	Decadal Growth Rates (%)
1951	4.43	27.23		62.4	17.29	
1961	5.31	25.74	19.64	78.9	18.0	26.44
1971	7.49	28.06	41.05	109.1	19.91	38.22
1981	10.60	31.10	41.52	159.5	23.70	46.23
1991	14.24	34.47	34.34	217.2	25.71	36.09
2001	18.93	37.36	32.94	285.4	27.78	21.35

Source: http://udd.gujarat.gov.in/Default_files/UrbanScenario.htm

2. See for more details on the project http://www.spcept.ac.in/pas_project.aspx?

In 2009, the share of the State's urban population in Municipal Corporations is estimated to be 58.4 per cent. The smaller Urban Local Bodies (ULBs) which are classified as municipalities are divided into four categories based on their population size and local importance. Of these the smaller Class B, C and D towns, though large in number, have only one-fourth of the total urban population in the state.

Table 2: Distribution of Urban Population in Different Classes of ULBs in Gujarat

Category of ULB	*Number of ULBs*	*Total Population in 2001 (in million)*	*% to Total Urban Population*	*Population in 2009 (in million)*	*% to Total Urban Population*
Municipal Corporation	7	9.7	57.5	14.5	58.4
Municipalities					
Class A (more than 100,000)	18	2.4	14.0	3.8	15.4
Class B (50,000 to 100,000)	33	2.1	12.4	2.9	11.5
Class C (25,000 to 50,000)	44	1.5	8.7	2.0	7.9
Class D (less than 25,000)	64	1.2	7.4	1.7	6.8
Total	**166**	**16.9**	**100.0**	**24.8**	**100.0**

Sources and Notes: 2001: Based on Census of India, 2001. It does not include the other 74 census towns with a population of 1.4 million which do not have urban local governments; 2009: projections using ratio method done in the PAS Project.

Gujarat has seen significant growth in income in the past few years. Its average annual per capita income at current prices in 2008-09, at Rs. 49,251 is 23 per cent higher than the national average, and shows an increase of 8.4 per cent over the previous year.[3] Ahmedabad and Surat, the two main cities in Gujarat, figure prominently in the top cities of India in terms of growth and average income. Surat's average household income in 2007-08 was next only to Mumbai's. "After adjusting for cost of living, the diamond hub of Surat is now the richest city in India. Between 2004-05 and 2007-08 Surat's middle class doubled in size and

3. Based on Directorate of Economics and Statistics (2010), "Key Statistics of Gujarat State 2009-10", Government of Gujarat, Gandhinagar, p. ix.

its low income category was reduced by a third (TRENDSNIFF, 2008)." Along with the rise in income, Gujarat has done well to reduce urban poverty over the past decades. From a high share of poverty in 1973-74 at 53 per cent, the share of the urban poor in total urban population has come down to 13 per cent in 2004-05 (the national average is 28.3%). Particularly, over the last decade there is significant decline in urban poverty in the state (MHUPA and UNDP, 2009).

Rapid rise in economic activities, rise in income and reduction in urban poverty has implications for drinking water demand. Underground water sources, the mainstay of drinking water supply in urban Gujarat, are overexploited. Narmada canal water has been a boon to ULBs in Gujarat in meeting the rising water demand.

Moving towards Water Security

Sources of Water Supply and Treatment: In North Gujarat, Saurashtra and Kachchh, available water resources are less than the requirements. In absence of perennial sources of surface water, ground water used to be the main source for drinking water. With overexploitation of groundwater in these regions, the water table was being depleted at the rate of 3 to 5 metres per year and water had to be abstracted from depth of 300 metres. Hirway (2005) states that 87 per cent of urban water supply was through groundwater and three-fourths of urban centres in Gujarat use groundwater (Hirway, 2005).

This situation has changed significantly since the Narmada canal based drinking water programme has been initiated. In 2009, only 35 per cent of the ULBs depended exclusively on ground water as a principal source of water, while 18 per cent are dependent exclusively on Narmada canal water. Nearly half of the ULBs use both Narmada canal water and groundwater.

Table 3: Source of Water Supply for ULBs (2009)

Source of Water Supply	*Number of ULBs*	*% of ULBs*
Bulk Water Purchase	30	18
Ground Water	59	35
Own Source (surface)	6	4
Mixed Sources	71	43
Total	**166**	**100**

Source: Based on surveys done under the PAS Project, 2009.

The Municipal Corporations, where nearly 60 per cent of the urban population of Gujarat reside, account for 73 per cent of the total municipal water supply (see Table 4). Among the water sources, bulk water purchase from Narmada accounts for 46 per cent of total water used by ULBs, while groundwater usage has dropped to 21 per cent. However, in terms of quantum of water supplied to the ULBs from the Narmada canal, the Municipal Corporations use 71 per cent of it. So far 72 ULBs receive water from the Narmada Canal, and another 19 have already been taken up.[4]

Table 4: Source of Water by Size Class of ULB (2009)

Category of ULB	*Own surface sources (MLD)*	*Ground water (MLD)*	*Bulk water purchase (MLD)*	*Total water (MLD)*
MC	1003.8	300.9	1072.0	2376.7 (72.9%)
A	20.5	86.3	215.9	322.7 (9.9%)
B	27.5	128.5	100.9	256.9 (7.9%)
C	14.8	81.4	63.9	160.2 (4.9%)
D	5.3	94.3	44.8	144.3 (4.4%)
Total (MLD)	**1071.9 (32.9%)**	**691.4 (21.2%)**	**1497.6 (45.9%)**	**3260.9**

Source: Calculated from PAS survey.
Note: Totals may not add up due to rounding.

Improving water security for ULBs dependent only on Groundwater: The 59 ULBs dependent only on ground water are mainly in central Gujarat and North Gujarat (see Table 5 for a full list of these ULBs). When the location of these ULBs is overlaid on the groundwater resource map of Gujarat (GWRDC) it shows that of these 59 ULBs, 10 ULBs are in critical and overexploited zones. These ten ULBs will need to be provided water from Narmada canal

4. Analysis based on information reported in Government of Gujarat (2010), "Statewide Water Supply Grid in Gujarat", Mimeo. Downloaded from http://www.gwssb.org/pdf/narmadaprojects.pdf

Table 5: List of Cities in Relation to Ground Water Zones

Ground Water Category	*Characteristics*	*No. of ULB*	*Name of ULB*
Over Exploited (red) >100	Restricted for Ground Water Extraction	5	Idar, Patan, Thara, Tharad, Dhanera
Critical (Dark) 90-100	Not Advisable for Ground Water Extraction	5	Dholka, Mehmadabad, Deesa, Manavadar, Vanthali
Semi-Critical (Grey) 70-90	Need Caution for Ground Water Retraction	7	Bantawa, Savarkundla, Bavla, Viramgam, Kathlal, Keshod, Visavadar
Safe (White) < 70	Feasible for Ground Water Extraction	39	Halol, Bayad, Khedbrahma, Dabhoi, Savli, Bagasra, Anand, Anklav, Boriavi, Borsad, Karamsad, Oad, Petlad, Sojitra, Umreth, Vallabh V. Nagar, Devgadh Bariya, Salaya, Talala, Una, Chaklasi, Dakor, Kanjari, Kheda, Mahudha, Nadiad, Thasra, Rajpipla, Vijalpore, Kaalol, Kutiyana, Prantij, Talod, Bardoli, Mandvi, Tarsadi, Chotila, Karjan, Umargam
Saline TDS >2500 ppm	Blending of Ground Water and Surface Water as per need	3	Bhabhar, Patdi, Harij

Source: Based on analysis in the PAS Project.

or other surface water source. There are 3 ULBs in high salinity region and they will need to shift to a mix of groundwater and surface water. All ULBs in Gujarat require undertaking serious efforts towards recharging ground water.

Access and Coverage of Water Supply: In 2001, about 91 per cent of urban population in Gujarat had access to piped water supply, exceeding the national average by almost 20 per cent and ranking among the first three states in India in terms of access to water services (Marie-Helene, 2006). Access to piped water in Gujarat was improved by about 18 per cent between 1991 and 2001 (Census of India, 2001). Results of the 58th round of NSS, 2002 provide a similar figure of 92 per cent of population of the state having access to piped water supply, compared to the national average of 74 per cent. It also reveals that only 7.5 per cent of households in urban Gujarat depended on groundwater sources (handpumps, tubewells or borewells). This is much lower than the national average of 25% households). NSSO (2002) also reported that 48 per cent urban households in Gujarat had drinking water facilities for their exclusive use (compared to 31% all-India average), 14 per cent have shared facilities and the balance 38 per cent use community facilities (compared to 55% all-India average).

The National Family Health Survey NFHS-3 conducted in 2005-06 and the District Level Health Survey DLHS-3 conducted in 2007-08 suggest a further improvement in coverage of water supply to about 95 per cent of urban population in Gujarat. The NFHS-3 suggests that 82 per cent of urban households in Gujarat have water taps in their house or in the yard. This is the highest among all states in the in the country (Mehta, 2010).

The PAS Household level surveys conducted in 2009 also found that 82.5 per cent urban households have access to municipal piped supply (with 63 per cent households having individual connection and others having shared connections). As compared to the NSS 2002 survey, this suggests a significant improvement in level of water service in Gujarat. The PAS survey results based on the information provided by ULBs, suggests that 68 per cent of households have water connection on their premises. Surprisingly, the variation across different size classes of cities is not very high. The Class A municipalities have a lower average than other classes of municipalities.

The PAS survey also collected information on access to water services of the slum dwellers. It revealed that while over 96 per cent of slum dwellers have access to water supply, nearly 53 per cent of slum households have a water connection on their premises. Again, the averages across different size classes of municipalities are quite similar. This suggests that in Gujarat many ULBs do not discriminate between a slum and a non-slum house in granting water connection.

Levels of Water Supply—Many ULBs Still Insecure: Table 6 provides information on key performance indicators for water supply in the state. The average water supply of 77 litres per capita per day (lpcd) in Gujarat is much lower than the state's goal of 100 lpcd and the norms laid down by WHO of 135 lpcd. There is a wide variation across the ULBs. From Table 7, we see that 26 ULBs provide less than 40 lpcd, which is the norm for rural water supply schemes. At the higher end of this spectrum, only 10 ULBs provide more than 135 lpcd.

Table 6: Average Service Level and Performance Indicators by Type of ULBs in Gujarat

	State Average —all ULBs	*Municipal Corporation*	*Municipalities*			
			Class A	*Class B*	*Class C*	*Class D*
Coverage of HH Services (%)						
Water supply connections	68	71	54	74	67	68
WS connections in slums	53	58	54	56	47	53
Levels of Service						
Per capita water supply per day (lpcd)	77	123	71	78	71	78
Number of hours of supply/day	1.5	1.2	2.1	1.4	1.5	1.6
Number of days of supply/month	23	25	24	22	22	22

Source: Based on PAS Survey.

Table 7: Per Capita Water Supply (LPCD)—Gujarat State

Category of ULBs	*Ranges of LPCD*					
	11-40	41-70	71-100	101-130	131-160	161-190
Corporations	0	1	1	1	3	1
Class A	1	8	9	0	0	0
Class B	7	7	9	10	0	0
Class C	8	13	16	5	1	1
Class D	10	22	13	12	4	3

Source: Based on PAS survey.

Thus, although Narmada canal has improved coverage and increased supply of water in ULBs, a large number of ULBs in Gujarat are still not secure in terms of their water requirement. With rapid growth of urban population, the quantum of water required from Narmada canal and other sources will have to increase manifold.

This low level of water supply is available to the residents only for a very short time. The average duration of supply in a day is 1.5 hours, and the average number of supply days in a month is 23. In some large cities, water is supplied only for 20 to 30 minutes in a day, and often every alternate day. The consequence of this is that urban residents have to incur large capital expenditure in constructing sumps and install pumps and overhead tanks to get water through the day. If the norm of 24 × 7 water supply (i.e. 24 hours of supply in a day for 7 days a week), is implemented, a large part of this private capital cost and operating cost would be reduced. However, this would require the ULBs to be efficient in their water supply operations.

In response to this situation, the Government of Gujarat has initiated a number of schemes under its new programme for Swarna Jayanti Mukhya Mantri Shaheri Vikas Yojana (SJMMSVY) to provide water through Narmada canal wherever possible and ensure adequate sources of water supply to meet the norms of 100 lpcd of water for all ULBs. Adequate funds have been allocated to meet these goals. While the Government of India has focussed on larger cities, the state government has balanced this by increasing allocation of funds to smaller municipalities.[5] Thus, over the next three to five years, Gujarat seems set to achieve the goal of adequate water supply along with full coverage of household access to piped water in all ULBs.

Improving Sustainability of Drinking Water Services

While the goals of adequacy and coverage of drinking water supply are likely to be met in Gujarat, there are issues of sustainability of these goals. We examine below the non-revenue water and cost recovery as two important aspects of sustainability of drinking water

5. See the analysis in PAS Project (2011, forthcoming), "UWSS sector finance and monitoring system: a baseline assessment".

services. These two parameters reflect drinking water management practices and have an important bearing on the overall service levels.

Extent of Non-Revenue Water (NRW): Non-Revenue Water is an important indicator for assessing the efficiency in service operation of a water supply system. NRW refers to water that has been produced but is "lost" before it reaches the customer. It refers to the amount of water produced that does not earn any revenues for the ULB. This "lost" water could be due to real losses (through leakages, also referred to as physical losses) or apparent losses (theft, illegal connections, free water etc.). High levels of NRW seriously affect the financial viability of water supply provision due to lost revenues, and increased operational costs adversely impacts the quality of the service provided.

Table 8: Estimated Non-Revenue Water among ULBs in Gujarat

Category of ULB	*Non-Revenue Water (%)*
Gujarat state Average	30
Municipal Corporations	34
Class A	33
Class B	26
Class C	32
Class D	30

Source: Based on PAS Survey.

The average NRW across all classes of ULBs in Gujarat ranges between 26-34 per cent indicating that nearly one-third of water is "lost" in distribution. The GoI 'benchmark' for NRW is 20 per cent (MOUD, 2008). The ULBs in Gujarat, due their inefficiencies are 'losing' nearly one-third of the water that they supply. If the NRW is reduced to the benchmark level of 20 per cent, there would be an increase in water supplied to the consumers and an increase in revenue to the ULBs.

It must be recognised that NRW is difficult to estimate in a context where there are no measurements of flow at source of water or at major distribution points and consumer points. In the PAS survey, the NRW measurements are based on estimates of water supply and consumption. Better measurement of NRW can be done by conducting preliminary water audits. Our estimates of NRW are on the lower side, as we have made assumptions about water consumption at the consumer end. No ULB in Gujarat has water meters, nor do most ULBs in Gujarat conduct water audit to assess the quantum of NRW. In one municipality in Gujarat, where a preliminary water audit was undertaken recently, the NRW was estimated to be 44 per cent, as against the 33 per cent value derived under various assumptions.

In order to gauge a better assessment of 'real' NRW, it is essential for all ULBs to undertake at least such preliminary water audits. This will inform ULBs about how much water is being lost, and why. Reduction of NRW can be achieved by improved management practices of detecting leaks and repairing them quickly, identifying illegal connections and collecting revenue from them, and regular monitoring of NRW by establishing District Metering Areas (DMAs) and installing meters at supply points and at consumer end.

Recovery of Operation and Maintenance Costs: Financial sustainability of a water supply system requires that ULBs cover their entire operation and maintenance costs. The extent of cost recovery for operation and maintenance (O&M) is defined as a percentage of total operating revenues from water supply-related charges to total operating and maintenance expenses on water supply. The total annual O&M expenses in water supply are estimated after excluding loan interest payment and depreciation. Cost recovery is estimated from the total annual operating revenues in water supply from local sources, excluding revenue grants. Such a 'full cost recovery' goal is also mandated for all ULBs that received grants under the JNNURM and UIDSSMT project of the central government.

Table 9: Percentage of Cost Recovery (O&M)

Category of ULB	*Range of O&M Cost Recovery (%)*				
	0-20	21-40	41-60	61-80	81-100
Municipal Corporations	1	1	2	0	2
Class A	3	7	3	3	0
Class B	3	8	10	3	3
Class C	8	13	10	4	2
Class D	11	13	13	9	5
Total	**26**	**42**	**38**	**19**	**12**
% of all ULBs	**19.0**	**(30.7)**	**(27.7)**	**(13.9)**	**(8.8)**

Source: *Based on PAS Survey.*

At the state level, the average cost recovery of O&M in water is 60 per cent. Nearly 50 per cent of the ULBs recover only 40 per cent of their operating costs, while 9 per cent recover more than 80 per cent of their operational cost. The variation across size class of ULBs is not much. On the whole ULBs in Gujarat recover only 60 per cent of their operating costs of water services. Even more disturbing statistics relates to collection efficiency. It suggests that ULBs in Gujarat collect only 50 per cent of the billed water charges.[6]

Table 10: Cost Recovery and Collection Efficiency in Service Delivery

Parameter	*State Average (all ULBs)*	*Municipal Corporation*	*Municipalities*			
			Class A	*Class B*	*Class C*	*Class D*
Percentage of O&M costs recovered through local taxes/charges	60	64	53	65	55	62
Collection efficiency of local taxes/charges for water supply (%)	50	51	53	51	54	45

Source: Based on PAS Survey.

An important component of Operational cost of water is cost of electricity. On an average, ULBs spend Rs. 2.4 per kilolitre on electricity charges related to water supply (electricity charges for operation of pumps at supply and distribution points). Compared to Rs. 4 to 6 per kl charged for Narmada water, the electricity costs of water supply account for a significant part of O & M cost of water.

The electricity charges account for 40 per cent of the total cost of water production at ULB level. For the municipal corporations, it is Rs. 3.2 per kl and constitutes over 50 per cent of its operational costs. Narmada water has made a significant difference to the operating costs. The average electricity cost for the 59 ULBs using only ground water is Rs. 2.64, as against Rs. 0.89 for cities using

6 In most ULBs in Gujarat, the water charges are collected as a part of annual property tax bill. This suggests that the overall revenue collection efficiency of ULBs in Gujarat is low.

Table 11: Cost of Electricity for Water Supply

Indicators	*Average for all MC*	*Average for Municipality*	*Average for all ULBs in the State*	*Average for ULBs using only Ground water*	*Average for ULBs using Narmada water*
Unit electricity cost (Rs. per kl)	3.17	2.35	2.40	2.64	0.89
Electricity cost as % of total O & M expenditure	53.0	39.5	40.2	45.4	20.9

Source: Based on PAS Survey.

only Narmada water. One municipality that uses a mix of Narmada water and ground water has a very high cost of water production of Rs. 15 per kl of which electricity charges account for 40 per cent.

Human Resources Capacity among ULBs: Another key dimension in sustainability of services in future requires addressing critical gaps in human resource capacities with ULBs. This is evident from the situation regarding availability of staff in Municipal Corporations and municipalities in Gujarat. In 2004 the urban development department had formed a special committee to review the minimum levels of staff required for different departments across municipalities by class.[7] The committee had specified the staff required for different types of ULBs. There are critical gaps in current human resource capacities available with ULBs. This is evident from the situation of many vacancies of sanctioned staff positions in Municipal Corporations and municipalities in Gujarat. For example, the PAS Survey of ULBs in 2008-09 showed that municipalities in general have only 50 to 60 per cent of the 'sanctioned' staff. Even of the available staff, a large proportion is non-technical and often temporary or daily wagers.

In 2007, GoG has created three common state cadres for: (i)

7. According to GR No: NPM/1089/1122/R, UDD, Gandhinagar, Government of Gujarat on Minimum Standard and Criteria for Municipal Staff to be maintained at ULB level dated 22-1-2004 directed the municipal bodies to maintain a minimum number of staff for different departments.

Municipal Accounts Officer, (ii) Municipal Health Officer and, (iii) Municipal Engineer, over and above the existing state cadre of Municipal Chief Officer.[8] Yet, in many ULBs the position of a municipal engineer remains vacant which adversely affects service delivery.

While the numbers and capacity of human resources within ULBs is important for planning and service delivery, it is equally important to ensure that staff has the necessary incentives for improved performance. This is more difficult and is linked to both administrative processes and institutional/structural issues. Some of these issues are covered in the administrative and structural reforms under JNNURM. The administrative reforms highlight ways to reward good performance, importance of strengthening internal systems and processes, as well as citizen interface. These reforms also refer to the use of standardised service level benchmarks (SSLB) to periodically measure and report on them. Ideally these service levels can be linked to staff performance assessments.

Financial Sustainability: One of the key challenges for ULBs in Gujarat is to make water supply financially sustainable. As we have seen, the cost of water supply in Gujarat is high and ULBs recover only about two-thirds of it from water tariffs and taxes. The usual response to such a situation is to raise tariff, but that is often not sufficient and solves only a part of the problem. What is required is to look at the entire gamut of operations and look for areas in which costs can be reduced and revenues enhanced.

As a demonstration of how this can be done, we look at the case of Kalol municipality. Kalol is located about 20 km by road from Gandhinagar about 30 km from Ahmedabad. It has a population of approximately 150,000 (111,700 as per 2001 Census). It receives its water from both surface and ground water sources. The major source is Pratappura Water Treatment Plant set up by the Gujarat Water Supply and Sewerage Board (GWSSB) that takes about 12.5 MLD of water from Narmada Canal. Besides this, the 21 boreholes within the city are used to draw about 5 MLD of water. The municipality has 17,800 connections serving about 30,000 households. With the present estimated population, the supply requirement is about 23 million (assuming 135 lpcd). Thus there is a shortfall of 6 MLD of water.

8. As per the GR No. NPM-102005/2054-R, UDD dated 28-9-2007.

A preliminary water audit suggests that the NRW in Kalol is 44 per cent, which is much higher than the 33 per cent reported by the ULB. It is thus possible to meet the shortfall of water in Kalol by reducing the NRW. Physical losses account for a major part of NRW and the municipality would benefit immensely from plugging all the leaks and monitoring flows in the system regularly.

Kalol municipality spends nearly Rs. 37.5 million a year to operate and maintain its water supply and collects only Rs. 830,000 a year. There is a large gap between its water related expenditures and corresponding receipt. As a first option, one can begin to look at ways to cut expenditure. If the physical losses are reduced (by plugging leaks), there will a saving of Rs. 6.4 million. Similarly, the energy costs can be reduced by Rs. 4.4 million through replacement of old and inefficient pumps at a few critical locations. On the revenue enhancement front, Kalol needs to improve its collection efficiency from 48 per cent to 90 per cent, identify and regularise illegal connections and start billing them and expand coverage in slums and in other areas. Our analysis shows that tariff revision should be the last option to be considered as this helps avoid passing on the ULB inefficiencies to its paying consumers.

Conclusions

Use of Narmada canal for drinking water has brought much needed relief to urban residents of Gujarat. There has been an increase in accessibility and better level of service in cities of Gujarat. With reduced dependence on ground water sources for drinking water, there has been a reduction in energy cost and rise in ground water table due to recharge.

However, it needs to be recognised that Gujarat is rapidly urbanising and the demand for water has been rising. With planned infrastructure investments in waterborne sewerage system, the water requirements in cities will increase significantly. The current levels of supply of 77 lpcd will need to be augmented to meet the international norm of 135 lpcd not only for the present, but also for the future population. While it is possible that Narmada canal water will be available for the foreseeable future to meet the water demand, it is necessary to take adequate measures to ensure sustainability of the urban drinking water security that Gujarat is on course for.

While the level of funding for water sector investments in Gujarat may be sufficient in the short run to achieve the immediate objectives and targets of 100 lpcd set by the state government, there is a question for long term sustainability of these investments. At least three issues deserve close attention to improve the chances of long term sustainability: the need for adequate expenditure on O&M for Urban Water Supply services, extent of cost recovery at least of O&M (recurrent) costs to be able to operate the services effectively and generate at least an adequate surplus for immediate capital investment requirements, and adequate human capacity in the ULBs for good consumer responsive service delivery. The issue of adequate staff is also linked to appropriate institutional arrangements that provide incentives and opportunities to ensure good performance in delivery of water and sanitation services.

For financial sustainability of services it is important to recognise that water is an important resource and must be adequately priced. Narmada water that is available to ULBs in Gujarat is priced between Rs. 4 to 6 per kilolitre for ULBs. However, ULBs do not charge adequately for this precious and limited resource. More importantly, nearly one-third of water is 'lost' in supply and only two-thirds of the operating cost is recovered. In order to sustain investments in drinking water, focus will also need to be on improved management and governance. For example, in order to reduce wastage of treated water through physical losses, the state government needs to make it mandatory for all ULBs need to conduct at least a preliminary water audit and carry out basic leakage management measures along with metering. Metering would also help in demand management as shown in a few places in India.[9] Similarly energy audits are required for all ULBs to bring down energy consumption through viable small projects. For reducing NRW, it is also essential to identify all consumers for billing and regularisation of illegal connections. Priority is also needed for adherence to service charters for consumers with a

9. While detailed studies are not available, anecdotal information from Navi Mumbai Municipal Corporation (Maharashtra) and Hubli-Dharwad Municipal Corporation (Karnataka) suggest that when backed by appropriate pricing consumer metering lowers water demand. Similar practice in a private residential colony in Ahemdabad also resulted in reduced water consumption and increased hours of supply.

responsive consumer grievance redressal system. For effective planning and maintenance all ULBs need to develop asset management systems.[10] Only then, it will be possible to ensure that security for urban drinking water achieved at great costs is sustained for the future generations.

---

10. Utilities and local governments in Australia and Canada have the best practices in asset management. While ULBs in Gujarat will take a long time to reach such levels, a start needs to be made. This also needs to be facilitated by supporting ULBs to first develop a good database on assets that will enable better maintenance and planning. Introduction of such practices will be essential to sustain the infrastructure created through the additional investments being made in the sector.

Part 7

Ways Forward

26

Social Limits and Economic Partnership Potentials in the SSP Command

R. Parthasarathy

Given the constraints of water resources in Gujarat, the state's achievements in agricultural production especially in the irrigated areas are significant (Parthasarathy, 2010). However, the experience in improving efficiencies in the irrigation systems indicates different dimensions of problems and failures. The Sardar Sarovar Project (SSP) had envisaged a unique method for water supply and management in its original design. This method by and large hinges on user participation as a cornerstone for efficient management of canal waters for irrigation.

In order to mitigate some of the problems in managing irrigation water, one of the institutional decisions taken even for the extant canals was transferring irrigation management to farmers under Participatory Irrigation Management (PIM). Besides addressing some of the management concerns, the aim was also to reduce recurrent expenditure by the government, improve operation and maintenance (O&M), increase productivity of water in agriculture and enhance environmental benefits associated with water use in the agriculture sector. Perhaps it sought to break the vicious cycle of low investment and inadequate O&M through a focused reform programme, which necessitated price reforms accompanied by changes in institutional framework including the formation of Water Users Associations (WUAs).

The objective of making such associations was twofold. One was to distribute water systematically and second was to improve the conditions of the field channels. A number of factors are considered to be instrumental in this shift in approach. Important among them are financial constraints faced by governments to run irrigation systems and a gradual deterioration of the irrigation infrastructure, which have resulted in a number of problems such as unreliability of water supply in terms of both quantity and timing, poor maintenance of irrigation infrastructure and inequity in water supply. Problems of recovery of water charges, bureaucratic indifference and indiscipline among farmers have also added to the current plight of canal irrigation management, especially in the existing systems. In order to eliminate the prevailing constraints in canal irrigation, a change in approach was felt necessary. In fact, while planning for SSP many of these constraints have been visualised and hence a perspective plan has been drawn (for details see, SSNNL, 1989).

Since the past few years the Narmada water has been irrigating the fields in parts of the 1.8 million hectare command area. In some parts of the area an NGO is also collaborating with SSNNL in construction and setting up of the participatory irrigation management programme. Unlike the existing irrigation schemes, in the case of Narmada the plan has been to form Village Service Areas (VSAs) and have WUAs manage the water. Further, it was also envisaged that sub-minors would be constructed by the farmer organisations and the field channels as well. It is time, therefore, to take stock of the translation of the planned objectives in the field into reality. To that end, this paper seeks to explore some of these aspects. An analysis of these and related aspects is presented with the help of field data.

The chapter is organised into five sections. The first section discusses what was originally planned for managing the water and its distribution.

The second section assesses the current situations in two distinct areas. Firstly, the results of a rapid assessment of 14 villages in the Phase I Command area are discussed. In these villages, the plan of irrigation development, especially construction, is being implemented by the Sardar Sarovar Narmada Nigam Limited (SSNNL) through private contractors. Second are the results, again based on a rapid appraisal in villages where irrigation development has been implemented through WUAs assisted by an NGO. There is a plethora of evidence showing the differences between department-led irrigation management programmes and those assisted by an NGO (Brewer *et al.*, 1999; Parthasarathy, 2000 and 2008a). Irrespective of how the programme is implemented, it is often pointed out that there are financial as well as scalar limitations in its implementation (Parthasarathy, 2000). Therefore, in managing a natural resource, one could argue that there need not be one or two managerial models but there could be a vector of management options. It is in this context, the option of private sector participation is discussed in this chapter.

The third section presents both theoretical and empirical experience of private sector involvement in irrigation management. Importantly, this section, besides exploring this possibility, also highlights the social limits to the private management option.

The fourth section specifically discusses alternative options by using the results of a 2006 study conducted by CEPT University—CRISIL (CEPT-CRISIL, 2006) exploring the possibilities of private investments in developing and managing the canal command area. The last section presents the summary.

I. Planning for Prosperity: The Original Plan

The discussion in this section is based on the agenda outlined in *Planning for Prosperity* (SSNNL, 1989). The large command area of Narmada encompasses a diversity of climates, precipitation patterns and topography; land forms, soil types and soil capacity; and subterranean water resources. There is also an evident diversity in the land holdings, infrastructure for social and economic development and even ethnic distribution of the population. Planning for irrigation of the command area had been carried out against this background. According to the plan, the whole of the command area was to be provided with organised field channels and field drains where the natural drainage channels were absent or inadequate. Land levelling was also contemplated in undulating areas having slopes greater than one percent; and command area roads were to connect important marketing, agro-processing and farming centres with special emphasis on providing the missing links.

In so far as water management was concerned, a heavy demand for Narmada waters was anticipated and therefore, the strategies included volumetric supply of water and rotational water supply as in Punjab and Haryana. It should be noted that this method of distribution of water would have been different from the *Shejpali* (demand based supply) system in vogue in the extant irrigation systems in Gujarat. There were other important features of water management for the SSP command. For instance:

- Localisation of paddy and perennial crops for more organised control and efficient water use;
- The distribution system to be flexible for use of technologies. Some of this aspect has been discussed in Joshi and Acharya (2011; this volume);
- Owing to the heterogeneity of the Narmada command, for the purpose of planning cropping pattern, the command area has been divided into 13 homogenous regions based on rainfall, topography, soils, groundwater and drainage conditions;
- For each region a set of feasible crop combinations and crop sequence based on agronomy was prepared. Importantly, crop sets for different regions were generated for 60 per cent water use efficiency and two levels of groundwater utilisation, low and high.

A major shift towards high value crops like paddy and wheat was envisaged as a result of the project (SSNNL, 1989). Besides the recommendation for cropping patterns, there was a considerable emphasis on water use efficiency. It was acknowledged that the efficiencies of existing irrigation schemes were low (see Vaidyanathan, 2006; Parthasarathy, 2010) due to various factors such as excess availability of water in relation to developed command, water rates based on crop and acreage, inefficient field irrigation practices, unlined channels etc. Some of the major features considered in the SSP plan to improve the efficiency are:

- Controlled delivery of water
- Water rates on volumetric base
- Lining of canals up to 8 hectare delivery point
- Discouraging perennial and summer irrigation through canal waters

- Intensified command area development programme for on-farm development
- Training of irrigation management personnel

II (a) Impact of Narmada Waters: Phase I Command

At the time of writing, the canal network has been completed to the extent of 19,000 km and 26 branch canals have been constructed. Through this canal system the first phase of network is in place and the irrigation potential developed is for 503,000 hectares (See, Map 1).

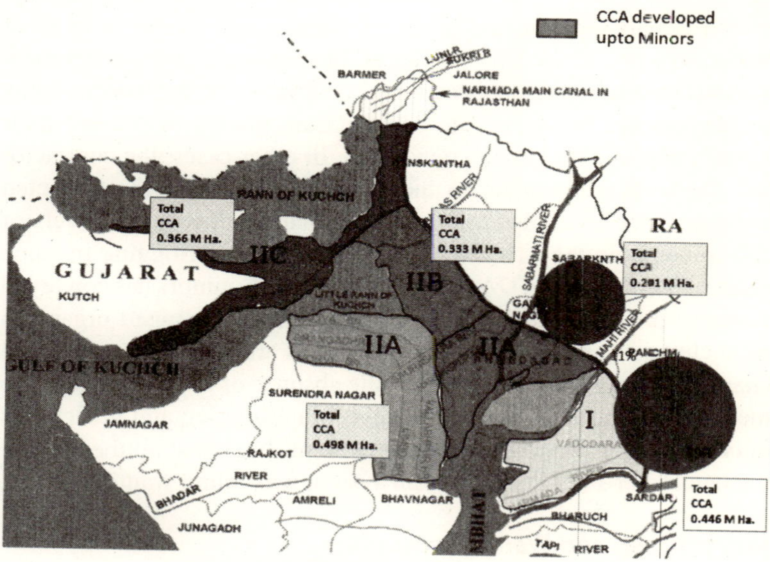

Map 1: SSP Command Area: Phases

To understand how the plans have been translated into reality, a study using rapid appraisal technique was conducted in 2010 in 16 WUAs of the first phase of implementation of SSP (see for details (Appendix Table I). Based on the list of WUAs that have management functions transferred, using random sampling procedure, five WUAs each were selected in two talukas—Naswadi in Vadodara district and Nandod in Narmada district, and six WUAs in Sankheda Taluka of Vadodara district. Besides a structured questionnaire to understand the details of quality of construction and benefits of the hierarchy of canals (minors, sub-minors etc.) under their purview, interviews were also conducted with presidents of the WUAs on the management aspects. The reason for selecting the WUAs of the first phase was that the construction of minors had already begun and WUAs had been formed. These WUAs fully or partially cover 77 villages.

Construction of Sub-Minors

In almost all villages there was concern about the quality of construction of sub-minors and the resultant constraints. Most of the sub-minors constructed 4-6 years back have started deteriorating, which is partly due to the practices farmers indulge in to avail of water. In some of the WUA villages, the construction of sub-minors was left incomplete either because the contractor left the work halfway or farmers did not allow the construction, mainly to prevent their lands being affected by the new canal. In other cases, the contractor left the work incomplete as there was a shortfall in the budgeted amount. There was also a reported mismatch between the levels of sub-minors and the irrigable land, leading to waterlogging in some places. In one of the WUAs none of the five sub-minors has conveyed water to the tail portions. While the plan had envisaged organised field channels and field drains, the farmers are yet to construct field channels in most of the WUAs. In the absence of proper conveyance mechanism, the problems of seepage, waterlogging and difficulties in conveying water to the field are evident. In almost all these cases the contractor-based construction system was found wanting.

Water Rates and Water Distribution

In the absence of fully developed conveyance mechanisms, volumetric supply of water could not be undertaken. As a result, flood irrigation is practised as in other irrigation schemes. In most of the farms, there was palpable overuse of water and the extent of waste water after irrigation often leads to conflicts among the farmers. In the Narmada command, the prescribed water rate is Rs. 328 per hectare per watering. Interestingly, four partly functional WUAs have voluntarily fixed water rates between Rs. 350 to 375 per watering per hectare. This difference in the amount collected is meant for meeting the WUA expenses, although they do not appear to be particularly enthusiastic about the task of visiting individual homes to collect water charges. As a result none of these WUAs could achieve 100 per cent collection, and consequently were not eligible for the 50 per cent rebate on the total water charges.

Usually farmers get water for their winter crops, and the rotation of water generally starts by the end of October. With a 15-20 days rotation period and an equally long gap, farmers normally get four to five waterings in a season. In the absence of set water distribution rules, they flood the fields, leading to shortage in the tail reaches. The practices include taking water by putting '*baknalis*' in the minor; piercing, underground pipeline, pump lifting, *heading* (blocking water) till their fields are irrigated and so on. Farmers use pipes to convey water to their fields located at 500-600 metres from the canal. These practices seem to set the social limit to water distribution as envisaged in the planning document of SSP. In turn, it also limits the effectiveness of SSNNL and the WUAs as after irrigation management transfer, maintenance O&M of canals has become the responsibility of the latter which none of them reported to have carried out save one.

In spite of the many issues discussed, of which most are at variance with the original planning, the changes in cropping pattern, the yield improvements and the secondary impacts in the study villages are noteworthy. A discussion on the findings of these aspects follows.

In the study villages, cultivation and animal husbandry are the major sources of livelihood. Before the advent of Narmada waters, farmers were heavily dependent on rain waters. Wells and tubewells were limited in number and some villages had small ponds or *kotar* for irrigation. In some predominantly tribal villages, dependence on monsoon crops also compelled the farmers to migrate seasonally to earn their livelihoods. With Narmada waters, the scenario began to change gradually in terms of cropping pattern, cropping intensity, use of farm inputs, farm practices and a substantial decline in seasonal migration.

Cropping Pattern and Yield

Table 1 reveals interesting findings about cropping pattern followed by the farmers before and after Narmada waters. The pre-SSP cropping pattern was mainly confined to *Kharif* (monsoon) crops, mainly of cotton, maize and *tur*. *Rabi* (winter) and summer cultivation was nominal and confined to only those farmers who had access to well or tubewell waters. After Narmada waters, cotton and tur remained prominent as *Kharif* crops with a phenomenal increase in area under cotton and maize in *Rabi* season. *Rabi* crops now also include wheat, groundnut and fodder. Summer cultivation was dominated by fodder crops, the others being bajri and groundnut. The Table also reveals an increase in area under cultivation in all the seasons. Apparently, access to Narmada waters had also inspired the farmers to level their lands, bringing fallow lands under cultivation. For example in Kotali minor, farmers have started growing grass in fallow lands. The change

Table 1: Cropping Pattern before and after Narmada

Sl. No.	*Crop*	*Area (ha)*			
		Before Narmada		*After Narmada*	
		Total	*Mean*	*Total*	*Mean*
Monsoon					
1.	Cotton	2710.4	180.7	3902.1	243.9
2.	Castor	40.2	13.4	199.5	28.5
3.	Tur	661.7	50.9	585.3	45.0
4.	Paddy	308.9	38.6	78.0	26.0
5.	Jowar	355.9	71.2	50.0	50.0
6.	Bajri	170.8	56.9	–	–
7.	Maize	696.5	69.7	383.1	47.887
8.	Mug	100.0	100.0	–	–
9.	Fodder	–	–	50.0	50.0
10.	Banana	–	–	5.0	5.0
	Total	5044.4	87.0	5253.0	105.1
Winter					
1.	Jowar	40.5	40.5	30.0	30.0
2.	Wheat	12.1	12.1	200.2	22.2
3.	Bajri	–	–	5.0	5.000
4.	Maize	–	–	1010.5	101.1
5.	Mug	–	–	57.0	19.0
6.	Fodder	–	–	75.2	25.1
7.	Groundnut	–	–	185.0	61.7
8.	Tal	–	–	2.0	2.0
9.	Chana	–	–	10.0	10.0
	Total	52.6	26.3	1574.9	49.2
Summer					
1.	Bajri	5.0	5.0	40.1	20.1
2.	Fodder	75.0	75.0	280.3	35.0
3.	Groundnut	–	–	75.0	37.5
	Total	80.0	40.000	395.4	33.0

in cropping pattern was also accompanied by adoption of HYV crops that have resulted in higher productivity.

Narmada waters have been instrumental in increasing the crop productivity (Table 2) for most of the crops cultivated by the WUAs studied. The yield of their major crop, cotton, more than doubled from 8.5 quintals per hectare to 18.2 quintals. Cultivation of maize in *Rabi* has also been more profitable as its yield has been much higher than that of *Kharif* maize. With most of crops showing an increasing trend in productivity, the overall crop productivity has gone

Table 2: Yield per Hectare (in Quintals)

Sl. No.	*Crop*	*Yield per hectare (Quintal)*			
		Before Narmada		*After Narmada*	
		Mean	*N*	*Mean*	*N*
Monsoon					
1.	Cotton	8.5	15	18.2	16
2.	Castor	17.5	3	25.5	7
3.	Tur	6.3	13	9.4	13
4.	Paddy	10.1	8	11.5	3
5.	Jowar	12.9	5	19.8	1
6.	Bajri	11.7	3	–	–
7.	Maize	11.4	10	15.3	8
8.	Mug	6.9	1	–	–
9.	Fodder				
10.	Banana	–	–	518.4	1
	Total	9.7	58	16.0*	48
Winter					
1.	Jowar	5.9	1	4.3	1
2.	Wheat	24.7	1	25.4	9
3.	Bajri	–	–	9.9	1
4.	Maize	–	–	30.8	10
5.	Mug	–	–	4.8	3
6.	Fodder	–	–	–	–
7.	Groundnut	–	–	14.7	3
8.	Tal	–	–	7.4	1
9.	Chana	–	–	4.9	1
	Total	15.3	2	21.4	29
Summer					
1.	Bajri	14.8	1	19.8	2
2.	Fodder	–	–	–	–
3.	Groundnut	–	–	19.5	2
	Total	14.8	1	19.6	4

* The average yield of banana has been excluded while calculating means for all *kharif* crops.

up from 9.7 quintals per hectare to 16.0 quintals in *Kharif*; from 15.3 quintals to 21.4 quintals in winter and, from 14.8 quintals to 19.6 quintals per hectare in the summer season.

Secondary Impacts

Animal husbandry is the complementary source of livelihood for most of the farm households. With Narmada waters, the farmers have an enhanced access to dry and green fodder and this has led to an increase in the number of milch animals, particularly buffaloes, and ultimately milk production. Besides the by-products of jowar, maize, paddy and wheat, farmers also cultivate *sundhiyo*—a green grass variety grown in winter and summer. In some of the villages like Handod, Kotali and Bhekhadiya, increased access to fodder has resulted in increase in income from animal husbandry by 10 per cent to 50 per cent. In some of the villages, such as Chametha and Suka, milk dairies have also come up in recent years. In Kothiya village where Gir buffaloes dominate the bovine population, the WUA president said that the increase in income from cultivation and animal husbandry has led to improved living standards.

Another impact of Narmada waters has increased groundwater level. The figures in Table 3 reveal that water level in wells has risen from about 43 feet to 29 feet and in tubewells it is 136 feet before Narmada waters to 93 feet now. Narmada waters have also helped in retention of water in summer. In fact there is an increase in the number of tubewells.

Table 3: Change in Underground Water Level

Sl. No.	*Taluka*	*Water level (in feet)*			
		Wells		*Tubewells*	
		Before	*After*	*Before*	*After*
1.	Sankheda	–	–	107.50	88.75
2.	Naswadi	51.67	41.67	106.67	70.00
3.	Nandod	37.50	20.00	177.00	111.00
	Total	43.57	29.29	136.25	93.33

It is apparent that the Narmada waters have led to multiple improvements even in the absence of implementation of the planned programme. It has been argued that involving NGOs would help in managing the social limits posed in proper management of the irrigation waters. In the section below, an exploratory assessment is provided of a group of WUAs that had the support of an NGO.

(b) Impact of Narmada Waters: WUAs with NGO

As part of the rapid appraisal of the performance of WUAs and water management in the SSP command, a study was also carried out in Vehlal village of Dahegam taluka in Ahmedabad district. Development Support Centre (DSC), an NGO is involved in implementing PIM in a command area of 5,000 hectares of the Vehlal branch that covers 12 minors and 13 villages. The implementation of PIM programme on the Vehlal branch was taken up as a part of the pilot project of SSNNL in 2004 and was completed in 2009. Even after the completion of the SSNNL funded project, the NGO continued to strengthen the 12 WUAs it had helped to form. There are sharp differences in the procedures and outcome of the initiatives with the involvement of the government and NGO in the implementation of PIM. The initial process adopted by both the government and the NGO was focussed on creating awareness about the programme among the farmers and building their capacity manage irrigation by themselves.

As the construction of sub-minors was supposed to be carried out by the WUAs in the Vehlal branch command area, DSC educated the farmers on the subject and the potential benefits of farmers' involvement in it. To facilitate this process, a Memorandum of Understanding (MoU) was signed between the WUAs and SSNNL. Overcoming the challenges and protests that came in the way of construction of sub-minors, the WUAs could accomplish the task with a total expenditure of Rs. 42.3 million up to September 2010 with technical support from the NGO and SSNNL engineers. Quality of work was ensured by timely supervision.

Another important achievement of the WUAs assisted by DSC is access to water. The basic objective of forming WUAs was to ensure equitable and efficient use of water. In the 11 DSC-supported WUAs of the Vehlal branch, technically sound and systematic construction of sub-minors was instrumental in increasing farmers' access to water in all reaches. The WUAs were also engaged in various other water distribution related activities. This is in sharp contrast to the other WUAs discussed earlier in places where the sub-minors were constructed by private contractors and there was no apparent role of the WUAs in water distribution. It is evident from the various facets of implementation that the government and an NGO could form a mutually beneficial partnership. For this to build up however, the PIM policy needs to be strengthened.

Meinzen-Dick *et al.* (1997), Brewer *et al.* (1999) and Parthasarathy *et al.* (2002, 2008) believed that the incremental returns that member farmers expect from the WUAs may be an important factor that influences the individual's decision to participate in WUA activities. Till now, no studies have successfully concluded that the formation of WUAs has led to an increase in agricultural productivity or revenue to the farmers. It is a well known fact that benefits potentially arise from adequate and timely supply of water or decreasing the cost of irrigation including conflicts. In absence of participation by member farmers and lack of WUAs' capacity to influence water supply, the possibility of degradation of water resources is inevitable in notionally WUA-managed schemes.

The second lacuna of PIM as implemented is WUAs inability to raise water charges (above those prescribed by the government). In order to instill financial self-sufficiency to WUAs, institutional directives have granted them powers to collect and raise water charges higher than the government rate. Very few WUAs have taken advantage of this provision.

Thirdly, WUAs are seen by both policy makers and irrigation professionals—mostly civil engineers—as a standard design for a new institutional infrastructure. This can be made clearer by the fact that before conducting any special repairs, farmers' organisations are required to obtain approval from the Irrigation Department. Evidently, such transfer has failed to encompass full control by the farmers' organisations over maintenance O&M. This eventually leads to their losing interest.

Fourthly, with this background, irrigation management transfer (IMT) to farmers may not be the optimal solution, especially in SSP where WUAs are located in isolated parts of the canal command and do not have horizontal or vertical linkages. In such a situation, management skills of WUAs in isolation cannot increase water use efficiency of the scheme as a whole. A federation of WUAs would make a difference to water use strategy and efficiency.

It is also evident that forming WUAs and tinkering with their managerial skills has so far failed to yield significant results at the scheme level in a scenario where on-farm efficiency of water utilisation is very low. Reform measures are, therefore, required. In the present paradigm, the components of water institutions are obviously influenced by strictly exogenous factors like supply of water. Since such factors define the context and represent the constraints as well as opportunities of the institution's performance, they play a major role in shaping both the nature and the character of the relationship between water institutions and water sector performance. However, partial emphasis on water institutions—water policy in general and water users association in particular—fails to bring about efficient utilisation of water resource. In a situation where the WUA faces constraints to take any legal action against its errant members or the department for not supplying agreed water quantity, efficient utilisation of water is an unthinkable proposition. If one goes by Ronald Coase's arguments, voluntary negotiation by the WUA (supported by, say, NGOs) and irrigation department for water management may in some circumstances lead to an effective outcome. This condition may arise when there is a well-defined property right. But the present reform agenda embodied in the PIM Act has its prime focus on water distribution, with no agency reform or policy formulation role for the whole system. In any case, presently water management is not considered a policy target. Such limited emphasis only on the formation of WUAs involves certain costs. These costs include finances required to ensure that the WUAs receive their due share of the water supply and to manage the irrigation system which has been transferred to them.

III. Possible Agenda for Development of SSP Water

The above analysis shows the role of the government and NGOs in providing a supportive and enabling institutional and policy framework for proper water use. There is another school of thought, including GoI (1995) that believes in a large potential gain of involving private players in water distribution and management. The National Water Policy (GoI, 2002) encourages greater participation by users in systems management and seeks to reorient water agencies toward improving the quality of service delivery by involving private parties. This policy was also strongly supported by the National Conference on PIM in 1995 (Hooja *et al.*, 2002). In order to ensure sustainable water use, the High Powered Committee Report (GoI, 1995) shows that the government is thinking positively about bringing private investments and managerial skills into surface irrigation. Ministry of Water Resources has justified the rationale of private sector participation by emphasising the benefits of the flow of private investments to the irrigation sector and has identified various incentives needed for promoting private investments especially in major and medium irrigation projects. The report also recognises the technical, legal, and institutional changes needed for promoting privatisation of irrigation.

The range of privatisation opportunities are expressed, both in analytic treatments and in policy declarations, in terms of the degree of ownership, control, management or operations that are predetermined by government (Hanke, 1987; Heath, 1989; Rondinelli and Kasarda, 1991). However, the terms "privatisation" and "divestiture" are often used interchangeably (Nankani, 1990; Luders, 1991; Cook and Kirkpatrick, 1988; Cowan, 1990; Berg and Shirley, 1987). Rondinelli and Kasarda (1991) offer a broad concept of privatisation, illustrating the various forms of privatisation measures: the sale of public assets to private investors, public-private partnerships, incentives/guarantees for private sector, transfer of public service delivery to NGOs, deregulation and liberalisation policies to facilitate private provision of public services, contracting out, government encouragement of private sector expansion in service industries, state subsidies for private service delivery, the reorganisation of an enterprise into separate entities, management and/or employee buyouts, debt-equity swaps, and build-operate-transfer (BOT) arrangements.

Countries like Mexico, Chile, Australia, USA, France and others have invited private parties to purchase concessionaries in water distribution and management. In India, options to involve the private corporate sector in project construction and management are currently under consideration in many states like Andhra Pradesh, Karnataka,

Madhya Pradesh, and Maharashtra. In most of these cases, the government was to acquire land and get all clearances for the project while the private company will complete the project following the operation plans and construction standards set by the government. Gujarat, besides focusing on farmers' participation on a large scale in irrigation management, was also looking at the viability of involving the private sector in mega-projects through joint ventures. Karnataka has taken steps especially in the construction of small and medium multi-functional projects and in the case of drinking water supplies where a public-private service model is being used. Madhya Pradesh was also planning to bring in private sector participation in projects where the dams are ready but not the canal networks (GoI, 1995; Saleth, 1999).

A unique and most notable attempt among the privatisation initiatives observed across the states was that of the Maharashtra Government inviting private bids for 52 irrigation projects worth Rs. 15 billion. These projects, with investments ranging from Rs. 10 million to Rs. 3.54 billion, were planned for allotment to private investors either on a build-own-sell basis or on a build-own-transfer basis. Besides the promise of an attractive return on their investment, private investors were also allowed the right both to promote fishery and tourism as well as to use 6 per cent of reservoir storage for promoting horticulture and floriculture. Private investors were also free from the hassles of land acquisition, environmental clearance, and R&R of displaced communities as the government would take care of those issues before handing over the projects to private operators (Saleth, 1999). Evidently, the basic philosophy of privatisation seems to be to improve water distribution and management. The literature attests that it is hard to predict the outcome of privatisation, but the form of privatisation pursued will largely determine the winners and losers of the new policy (Luders, 1991; Fitch, 1988; Domanski, 1992). Implementation success also depends on the management of the privatisation process. However, the effective assessment of the constraints and possibilities of privatisation necessitates a clear picture of the economic sectors that may gain or lose. The gains and losses are largely determined by how clearly the user rights are defined.

In SSP, where innovations have taken place in almost all conceivable stages, it is not impossible to set new standards in water management. In the circumstances where the Irrigation Department determines supply, water right is believed to bridge this link. Water rights will be a function of water available in the reservoir. Within WUAs, an entitlement system[1] can be conceptualised as a quota system. Water quota to the farmer members will be allocated on the basis of prescribed cropping pattern by the department and the share the household owns which again is determined by the size of land holding owned in the command area. This will leave farmers free to grow any crops, and sell water. Within each WUA, farmers will have water rights proportionate to their farm size. This system would lead to efficiently utilising water and would eventually transfer water from low to high valued uses. Rosegrant *et al.* (1994), Bruns (2000), Vermillion (1997) and Johnson (1997) argue that establishment of water rights provides incentive for water users to improve efficient utilisation of water.

In the present scenario, in which water supply is fixed and is determined by government, a revolution in irrigation agriculture is required which will link this given supply to ever increasing demand for water. If water rights—and that too tradable water rights—are vested in WUAs, which is not the case at present, it will provide economic incentives to maximise the net benefits generated by increasing production efficiency. This property induced cohesion is important in many aspects of water management but it is especially critical for water allocation which will enhance farmer members' confidence in services provided by WUAs and create an incentive for them to invest in the long term capability of system. Such a right will lead to greater involvement of users in repair and rehabilitation decisions thereby increasing the efficiency of water use rather than other approaches like initiating motivation, training etc., which is the case at present. In water scarce regions, if tradable water rights are established, water users can buy and sell water, which enables users to support the efforts of reallocating water.

For a better understanding of PPP opportunities and risks in the irrigation sector, an analytical framework is necessary. This is expected to have two necessary sets of functions: (a) the investment functions, including project financing, its design and implementation and the

1. The rules and procedures of physical division of a water resource between potential users.

governance functions of regulation and control, including water allocation, monitoring and supervision of irrigation management; and, (b) the O&M and management of distribution.

The decision to invest is evidently dependent on the returns on investment, which could be directly from user-fee or a combination of various commercial activities and government subsidies. A large irrigation development can involve a huge outlay and evidently the length of the payback period is long. The political and social risks associated with investment are recognised but kept out of the purview of the present analysis. One of the important functions of the new management is the water distribution function which requires a legal framework of water rights and a regulatory function to ensure that rights are protected. Equitable sharing of water among farmers is one of the challenges of irrigation management. It requires technical and economic skills to assess water demand against the available water resources. The negotiating skills to arbitrate any conflicts could be carried out by the WUA and hence this constituent of PIM is very important. Further, the O&M functions also entail management of water at the field level including assessment of water demand. This indeed provides the basic framework in the management or governance mechanism towards an economic understanding of how water functions as an input to farming. Some of the water distribution related functions like, monitoring and operation of gates for delivery of the water and billing, collecting water charges fees and management of contracts are also essentially functions of the WUAs with supervision by the investors. However, it is assumed that the crucial main canal maintenance is the government's responsibility, and so also coordination with the concerned agencies.

IV. PPP Options in Narmada Command

The earlier sections in this chapter discussed the government-and NGO-led models of setting up WUAs, monitoring of canal construction and management of water. A third model, Public Private Partnership (PPP) also seems possible in the array of available management options. A PPP model was visualised in building and managing the distribution network in the SSP command with an idea that it would be able to take care of a substantial future expenditure on building and operating the irrigation network. The scope of PPP would then include financing and constructing the canals, their O&M and water distribution. A pilot study towards assessing the feasibility of this model was undertaken during the year 2006 in the SSP command by CEPT and CRISIL[2]. The objective of the study was to assess changes in agricultural characteristics to ascertain what role these could have on the success of PPP; to determine the optimum structure for PPP in a pilot region; and, to assess the willingness and ability of the local institutions to participate in the process.

To understand the PPP possibility, Motibhoyan distributory which off-takes from the Sanand Branch Canal (SBC) at a distance of 1.44 km from NMC, was selected for the pilot study (Figure 1). According to the plan, the distributory would irrigate a cultivable command area of about 1,700 ha covering five villages of Kalol Taluka, Gandhinagar District, namely Wansjada, Ramnagar, Sherisa, Hajipur, Motibhoyan and one village of Kadi Taluka, Mehasana District—Wamaj. SBC was selected because the development of its command area was to be taken up in the second phase and planning for it was then in progress. First, the plan of construction of the minors was obtained

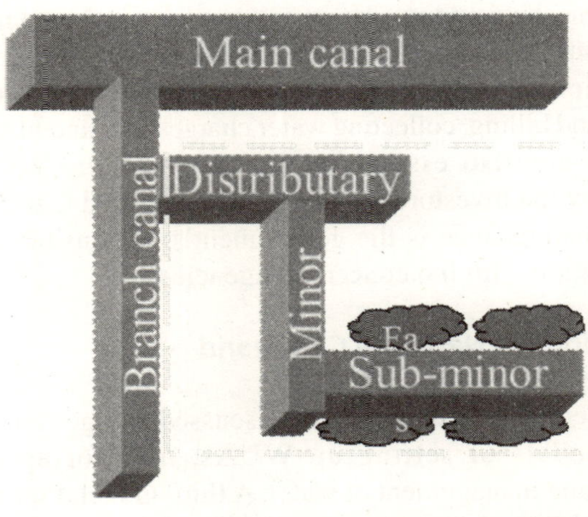

Figure 1: Schematic Representation of the Distributaries, Minor and Sub-Minor

2. CEPT University, Ahmedabad and CRISIL (2006): *Pilot Study for Public Private Participation in SSP,* a report submitted to the SSNNL.

from SSNNL. It showed a network of one distributory, five minors and 35 sub-minors. With the help of engineers the total cost of construction of this network, including the distributory with five minors 35 sub-minors, was worked out. At 2006 prices this came to Rs. 5.63 million (Table 4).

Four villages were selected to understand the possible changes in agriculture due to the imminent arrival of Narmada waters covering 600 hectares and 418 potential beneficiary households (Figure 2). A list of beneficiaries was first obtained from the report of Multimedia consultants (MC) that had earlier conducted a baseline study of the plots owned by individuals in the planned command area. While MC's report provided plot-wise information, the study had households as the unit of analysis. The details are provided in Table 5. While the benefiting plots due to Narmada irrigation are 1,342 (according to MC) the number of beneficiary households estimated in 2006 was 418. Table 6 indicates the number of households and plots of land in the study villages. A three-pronged approach was adopted to arrive at the number of potential beneficiary households from the plot-wise information of MC. First was the

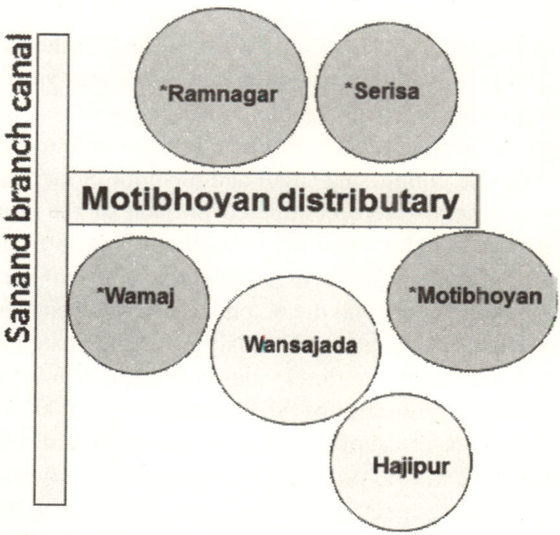

Figure 2: Schematic Representation of Study Villages in Motibhoyan Distributory

Table 4: Estimates of Constructing Distribution Network

Canal Type	No.	Length (km)	Cost (in million Rs.)					
			Land Acquired (in Ha)	Earth-work	Lining	Structures	Others*	Total
Distributory	1	3.77	0.17	0.36	0.34	0.28	0.46	1.61
Minors	5	14.20	0.40	0.89	0.57	0.29	0.004	2.15
Sub-minors	35	39.80	–	0.41	1.29	0.17	–	1.88
Total	–	–	0.57	1.66	2.20	0.74	0.46	5.635

Note: The break-up of cost may not add up to the Total due to rounding off error. The Total reflects correct estimate to the second decimal.

* Other construction costs cover construction cost for Approach and WBM Service road.

identification of *falias* or *vaas* and a list of land owners with plots of land in the command area was prepared. This was verified with the *Sarpanch* and/or *Talati* and later confirmed with each household. Interviews with potential beneficiary households and focus group discussions—one comprising small and medium farmers and the other with large farmers—were conducted in each of the study villages. Importantly, the characteristics of production pattern and expected changes in cultivation practices were studied in two scenarios—the current irrigation pattern and the potential pattern when Narmada water became available. The latter, obviously, was anticipated outcomes as projected by the farmers of these villages situated near NMC.

The potential benefits of SSP were found in all areas. These included changes in cropping pattern, incomes and willingness to adopt better agricultural practices those prevalent. Overall, a marked improvement was expected in cropping as well as irrigation intensity after the advent of Narmada water. More than 63 per cent increase in the Gross Cropped Area from 1187.11 ha to 1845.05 ha was envisaged

Table 5: Details of Minors, Beneficiaries and Survey Households

Name of Minor	*Name of villages*	*No. of plots to be benefited (as given in Multimedia Consultants' report)*	*No. of benefiting households*
A. Motibhoyan minor		801	
	Motibhoyan	539	151
	Sherisa	193	106
	Ramnagar	63	54
	Wamaj	8	
B. Wamaj Minor		541	
	Wamaj	541	107
			418
Total (A+B)		1342	418

Note: 8 beneficiaries of Wamaj village were residing in Motibhoyan village, hence included in the list of latter village.

Table 6: Distribution of Plots Owned by Households

No. of plots	*Villages*				
	Motibhoyan	*Sherisa*	*Ramnagar*	*Wamaj*	*Total*
1	466	153	51	466	1136
2	28	17	10	33	88
3	17	14	1	22	54
4	7	2	1	10	20
5	4	3		10	17
6	8	3		4	15
7	4			2	6
8	4			1	5
12	0			1	1
13	1				1
25	0	1			1
Total Plots	539	193	63	549	1344
Study households	151	106	54	107	418

Source: Multimedia Consultants Report, SSNNL, 2006.

in these villages as a result of the canal. The Gross Irrigated Area was expected to rise by 72 per cent from 840.51 ha to 1166.89 ha. The gross income of the households was expected to rise by almost 50 per cent. The difference in outcomes before and after Narmada water could also be seen as estimated by the farmers in terms of yield of major crops, which was expected to be twice the prevailing level. There is also an expected shift to high value crops. In the current situation, vegetables followed by *bajri* takes the prominent position whereas after Narmada water, paddy, wheat and *bajri* are expected to take significant positions (Figures 3 and 4). In 2006 there was limited awareness about drip and sprinkler sets, and market and credit opportunities were reportedly very limited. The formal institutional credit support was found to be insignificant and the farmers reported problems in getting input resources. The opinion of the households on even sale of produce changed significantly from the then scenario to the future scenario after Narmada; Around 90 per cent of them predicted an increase in the sale of produce. This is important in the current situation where the agriculture sector has been facing serious issues with respect to forward and backward linkages and the condition is dispiriting for the small producers. In such a context, it was expected that the WUAs would be able to streamline these issues with the organisational support that would enhance the collective bargaining power apart from making the service delivery feasible.

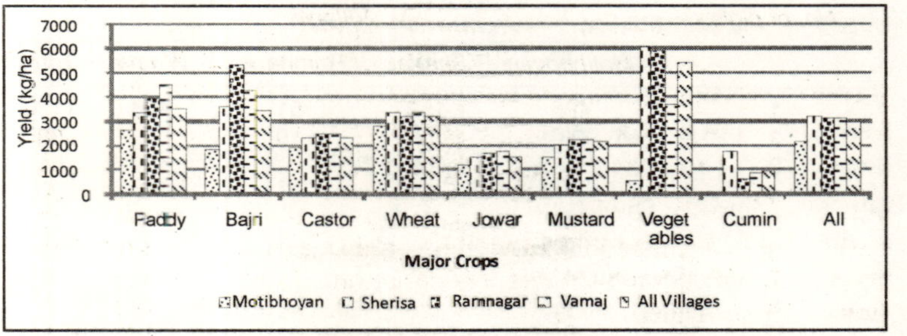

Figure 3: Yield of Major Crops in 2006

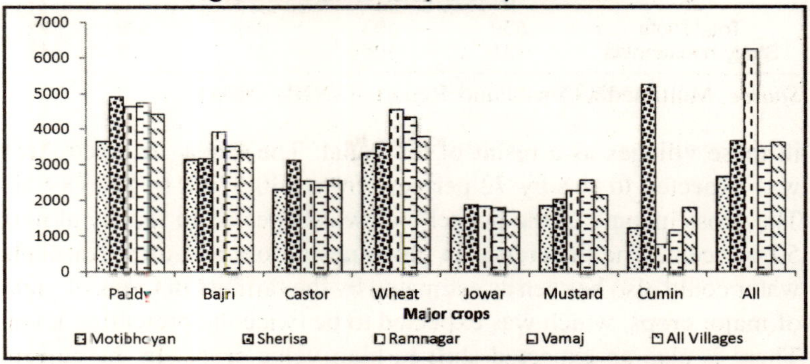

Figure 4: Expected Yield of Major Crops after Narmada Situation

The responses of households also related to willingness to contribute and abide by the prescribed agricultural practices and regulations after the advent of Narmada water. When asked about their opinion on paying water charges, a negligible number expressed refusal to pay. The largest proportion was willing to pay between Rs. 300-500 per season. On the proposition of irrigation water management by the private operator, a majority preferred to rely on the Government whereas less than five per cent showed concern over the profit motive of private investors. However, there were also many who thought that private operators could perform better due to competitive pressure and hence respond to the farmers' needs.

An analysis on the performance of irrigation before and after Narmada water showed that in situations of diverse interest of farmers and the Irrigation Department, no single measure of reform is expected to work. As water is one of the inputs used in cultivation, reforms in water distribution and use should be complemented by reforms in land regulation, existing infrastructure, market and credit facilities. There were possibilities of economic diversification at the household level, both due to non-agricultural employment and improvements in income due to Narmada water. Currently however, the prospects of growth are marred by poor infrastructure and weak extension services. In the context of efficient water supply, a well-functioning WUA could contribute positively in countering these shortcomings provided that they are federated and supported by a strictly implemented regulatory framework.

In so far as the potential investor is concerned, during discussions with farmers, irrigation department engineers, grain sellers, and others, the Agricultural Produce Marketing Cooperative (APMC) stood out to be the best placed potential private investor who not only could monitor construction but also manage water distribution through farmer organisations. APMC and the WUAs could obviously have mutual business interests in developing the network. Interviews with APMC office-bearers revealed that the benefits could be at least threefold. It was estimated that the APMC would earn around Rs. 200,000 per annum more from a higher volume of trade. It was also understood that the costs could partially be met through a surcharge on production or on irrigation that equals up to 5 per cent of the additional value, and by increasing the APMC's commission rate from members to 2 per cent. It was also felt that government could provide some subsidy partly from the savings on electricity that is now being used for deep tubewells. Monetisation of electricity savings would be substantial. APMCs reserves could be used for investing in this partnership and needed funds could be raised as loans from the banks. Being a cooperative, its profit expectations are long-term in nature and they would be invested only to improve the agricultural sector in the region. Figure 5 shows the APMC model.

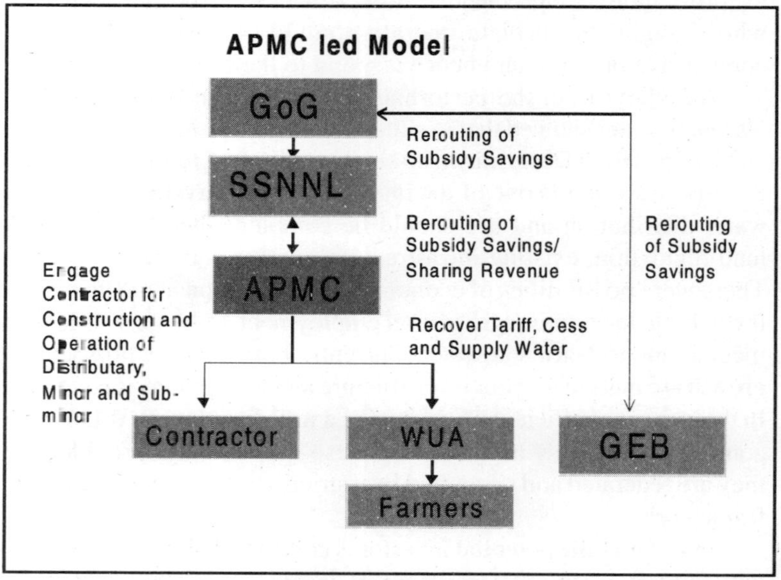

Figure 5: PPP in SSP—Potential Partner Management Model

Alternatively, if the APMC is not interested, private organisations may be sought to play the role of investor, though it would involve annuities to reduce revenue risk.[3] The government could underwrite at least to the extent the value of reduced electricity consumption, which could save Rs. 850,000 as estimated from the pilot area alone. The benefit through trade was estimated to be 1 per cent of additional

3. For details of the proposed mechanism see the report *Pilot Study for Public Private Participation in SSP (2006)*, prepared by CEPT University, Ahmedabad and CRISIL submitted to the SSNNL.

agricultural production. Further, if we assume that the required revenue is 15 per cent of Internal Rate of Return (IRR), this could be met by selling fishing rights in NMC which would contribute an additional financial IRR of 1 per cent. These benefits, if tapped strategically, would indeed make the PPP financially viable.

Overall, there are some important lessons learnt. First, it is possible to have alternative models to government financing and management of the canal network. APMCs may not be the first choice in all areas due to low financial reserves or low level of trade income. However, strong cooperatives and others like Gujarat State Fertilizer Company (GSFC) or Gujarat Narmada Fertilizer Company (GNFC) could be alternatives to APMCs in this model.

V. Conclusions

It is evident that the water management aspect has been meticulously planned for the SSP and the construction of minors and sub-minors was envisaged to be carried out by the users. Since the first phase of canal construction was completed and over 500,000 ha of command area is already irrigated, an attempt was made in this paper to understand the present status in the newly developed command area of SSP. It became apparent that in some places an NGO is also collaborating with SSNNL in setting up the irrigation management programme. Therefore, both these areas were studied to understand differences in implementation. In the NGO-assisted WUA, construction and water management was found to be better than in the contractor-led construction programme. Similarly, in areas where the NGO is present, water management practices, including that of fee collection, appear to be better. However, in both the areas the dividend due to irrigation waters is amply evident. For example, there are cropping pattern changes, improvement in yields and positive secondary impacts like rise in ground water level and so on. It should, however, be noted that with the advent of new waters, changes in cropping pattern and improvements in yield by default would be a self-directed function.

Since the success of an infrastructure investment is not determined by planning alone it is pertinent to examine the recent lessons learned from experience and their implications for future. It is apparent that there are many competing demands for SSP water; also that the irrigation network development costs are rising. Therefore, the future of SSP water management has to focus on both and take into account new developments. One of the possible avenues seems to be involving private players with well defined water rights. While public irrigation service tends to be supply-driven, private irrigation could be demand-driven which might also reflect the financial objectives. These characteristics, however, impinge on many of the basic decisions for development planning and imply fundamental influences on the investment approach. Whichever the model one pursues, the analysis highlights that community involvement at local level is a necessary condition; even as options at higher level are open to various players. For instance, the possibility of private operators in Sardar Sarovar Command even by large cooperative federations like APMC managing the distributories does not preclude the options to have effective water users associations below the outlet levels. There is in fact a social limit in all economic partnerships.

Acknowledgements: The author is grateful to Mr. Harish Joshi and Ms. Mahima Gupta for ably assisting him in the fieldwork and in analysis; Development Support Centre and staff; Ms. Jharna Pathak for her assistance in conducting the CEPT-CRISIL study in 2006, and to CRISIL for the excellent partnership in carrying out the study; and, to Ms. Aparna for her meticulous help in the preparation of this paper. I am also thankful to Professor Ravindra Dholakia for his comments on this paper and to Mr. Vivek Khadpekar for editorial support.

Appendix

Table I: Features of WUAs Visited in Narmada Command

Village/WUA	*Command area (CA)*	**Sub-minors fully constructed Yes-1, No-2, No sub-minors, only filed channels-3*	*% of CA irrigated by Narmada waters*	*Changes in cropping pattern after Narmada*	*Crop yield (quintal per hectare)*		*Rate of recovery (%)*	*Status of WUA F/PF/NF*
					Before	*After*		
Sankheda								
Dholi	311.0	2	16.08	Area under cotton and castor increased	8.67	27.17	40	PF
Handod	65.0	1	100.00	Change in variety—from shanker cotton to BT	14.40	33.10	NR	NF
Bihora	607.0	1	13.34	Increase in area under cotton/castor introduced/increase in area under wheat/summer cultivation of bajri and sundhiyu	15.93	18.30	NR	PF
Kotali	766.0	3	50.00	Started cultivating castor and area under cotton increased, replacing paddy/area under lucern and bajri increased in summer	14.34	20.76	NR	NF
Aritha	243.0	1	50.21	Area under cotton	17.28	18.90	NR	NF

Village/WUA	Command area (CA)	*Sub-minors fully constructed Yes-1, No-2, No sub-minors, only filed channels-3	% of CA irrigated by Narmada waters	Changes in cropping pattern after Narmada	Crop yield (quintal per hectare)		Rate of recovery (%)	Status of WUA F/PF/NF
					Before	After		
				increased/area under sundhiyu increased in winter/				
Vasan	306.0	2	9.80	Area under cotton and tur increased/Paddy, jowar and bajri discontinued/no crop in winter and summer	8.03	7.40	NR	NF
Naswadi Chametha	372.0	1	37.90	Area under cotton increased/castor introduced/cultivation of maize shifted from kharif to rabi season/groundnut introduced in rabi/	3.68	16.29	50	F
Amroli	336.0	1	.00	Area under cotton increased/un irrigated maize in winter	13.20	26.37	NR	NF
Kothiya	121.5	1	66.67	Jowar replaced by maize in rabi season/Sundhiyu in summer	6.90	16.47	NR	NF
Anandpuri-2	307.0	2	32.57	Area under cotton and tur increased/paddy	5.53	26.53	NR	NF

Village/WUA	Command area (CA)	*Sub-minors fully constructed Yes-1, No-2, No sub-minors, only filed channels-3	% of CA irrigated by Narmada waters	Changes in cropping pattern after Narmada	Crop yield (quintal per hectare)		Rate of recovery (%)	Status of WUA F/PF/NF
					Before	After		
				discontinued/maize cultivation shifted from kharif to rabi/whaet in winter/and sundhiyu in summer				
Piplaj	558.0	2	71.68	Area under cotton increased replacing bajri and jowar/castor in kharif and maize in winter and groundnut and sundhiyu in summer started	10.38	25.88	NR	NF
Nandod Mithivav	530.0	2	75.47	Area under cotton and tur increased/paddy discontinued/cultivation of maize and wheat in rabi and of groundnut in summer	7.30	12.30	NR 8	PF NF
Kharagada	468.0	2	42.74	Wheat and maize in rabi introduced	5.18	11.77	NR	PF
Bhekhadia	301.0	1	33.22	Cultivation of maize, wheat, mug, bajri and sesame in rabi and sundhiyu in summer	5.85	8.23	15	PF

Village/WUA	*Command area (CA)*	**Sub-minors fully constructed Yes-1, No-2, No sub-minors, only filed channels-3*	*% of CA irrigated by Narmada waters*	*Changes in cropping pattern after Narmada*	*Crop yield (quintal per hectare)*		*Rate of recovery (%)*	*Status of WUA F/PF/NF*
					Before	*After*		
Suka	236.0	1	42.37	Cultivation of wheat, mug and groundnut in rabi and sundhiyu in summer	13.53	16.10	15	PF
Vaviala	493.0	2	60.85	Cultivation of maize, wheat, gram and groundnut in rabi and sundhiyu in summer	6.27	18.23	10	PF
All WUAs	6020.5	–	43.93	–	9.98	18.13		–

Note: Poor construction of sub-minors was reported by most of the WUA representatives.
NR: Not reported as they had no idea.
F—Functional, PF—Partly Functional, NF—Not Functional.

27

A Case for Pipelining Water Distribution in the Narmada Irrigation System in Gujarat, India

Tushaar Shah, Sunderrajan Krishnan, Hemant P., Shilp Verma, A. Chandra and *C. Sudhir*

Introduction

The Sardar Sarovar Project (SSP) on the river Narmada, arguably the lifeline of the Indian state of Gujarat, has always been in the limelight for one reason or another (see figure 1 for a map showing the SSP command). Now, the project has got drawn into an altogether new controversy. Farmers in the SSP command have stolidly resisted the idea of giving up any land for the construction of the distribution system. As a result, while the Sardar dam is full, and it has enough water even with its present height to irrigate 900,000 ha, the actual area under gravity flow irrigation has reached around 75,000 ha, just over 5 per cent of the ultimate design command area. Given the persistent problems in land acquisition, it will take years to achieve targeted irrigation command, if at all. Frustrated by delays in laying the 66,000 km of distribution canal network to achieve the targeted 1.8 million hectares (ha) of irrigation, it was proposed that the government lay an underground pipeline network 3 metres below ground level. The key advantage of pipelining is that it avoids having to acquire land for surface distribution system.

The shaded area in the figure shows the command area of SSP in Gujarat.

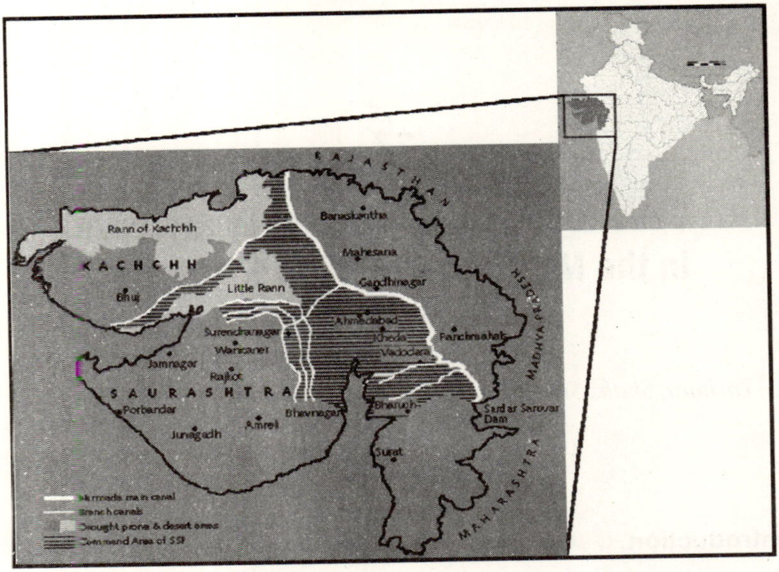

Figure 1: Sardar Sarovar Irrigation System in Gujarat, India

Source: (http://www.narmada.org/maps/map.gif); (http://upload.wikimedia.org/wikipedia/commons/thumb/0/01/India_Gujarat_locator_map.svg/543px-India_Gujarat_locator_map.svg.png)

The pipelining idea has, however, run into bitter criticism from SSP's master planners including C.C. Patel (*Times of India*, 2009a), one of India's best known irrigation engineers of the 1970s. SSP's planners admit that pipelining was considered during the late 1970s but rejected as 'unviable'.

This chapter argues that pipelining will fail under government construction and management but work very well if constructed and managed by farmers. It also argues that the critique of pipelining overlooks the vast changes that have swept Gujarat's irrigation economy since the 1970s when the SSP was planned. Many arguments against pipelining are today open to question. For example, it is argued that pipelining the SSP will mount Gujarat's farm power bill. However, if piped distribution of SSP water can substitute vast amounts of groundwater pumped from deep tubewells, pipelining can actually reduce Gujarat's total farm power bill. SSP's planners suggest that Gujarat should replicate irrigation institutions like *warabandi* (rotational water distribution) from the Punjab and Haryana experience. However, even with *warabandi* in canal irrigation, the significance of gravity flow irrigation has declined in Punjab and Haryana. Today these two states depend upon groundwater pumping for over 75 per cent of their irrigation (Dharmadhikary, 2005; Singh, 2006 citing a Government of Punjab 2005 document). Finally, SSP planners criticise pipelining as an impractical idea nowhere practised in the world for irrigation commands. The truth is that even within India, according to the Minor Irrigation Census of 2001-02, well over 2 million ha of land were irrigated by buried pipe distribution of surface or ground water (Government of India, 2005). Retrofitting canal systems by buried pipeline networks has also been extensively practised as part of modernisation of irrigation systems in advanced countries such as the United States, Spain and Canada (Shah, 2009).

Assessment of benefits and costs of pipelining critically depends upon the alternative it is compared with. When we compare pipelines on SSP with surface distribution systems, pipelining would naturally appear power-intensive. However, given its size, the SSP must be viewed against the larger canvas of the entire irrigation economy of Gujarat, which pipelining can alter in fundamentals ways. The key issue is who constructs and manages the pipeline systems: if the task is performed by government departments, chances are high that it will fail, given the dismal record of government managed pipeline schemes for irrigation dependent on gravity flow. But if the government encourages farmers and their co-operatives to construct and manage pump-based piped distribution, results can be successful as the Maharashtra example suggests. This chapter is a 'thought piece' written to generate a debate and facilitate more detailed analyses and calculation. We do not have the information needed to tie up all the loose ends of the argument, test all the assumptions implied and present complete cost-benefits of this proposal. However, the intent is to show that intuitively the pipelining idea has many advantages and it can throw up possibilities for institutional innovation and productivity gains that need to be factored into any serious analysis of this alternative.

Results

We outline at least 10 reasons why Gujarat should seriously consider pipelining the SSP distribution system (as well as other canal irrigation projects), and briefly describe the Maharashtra example that inspired our case.

Plan A is not working: SSP planners advocate proceeding with the Plan A of the SSP under which: (a) the Sardar Sarovar Narmada Nigam Ltd (SSNNL) would construct the main canal, the branch canals and a lined distribution system going up to the Village Service Area (VSA) of 250-400 ha; (b) within the VSA, the irrigators, organised into a Water User Association (WUA), would voluntarily contribute land, labour and other resources (including funds) to construct the network of sub-minors and field channels to distribute water within the VSA; (c) the WUAs would take the responsibility of orderly distribution of irrigation water within the VSA, ensure regular maintenance and repair, and collect water fees of which around 12 per cent would be reimbursed to the WUA as a managerial subsidy; (d) sophisticated, computerised water control through control rooms established along branches and distributaries throughout the command would ensure volumetric supply and charging for water.

Five years after the SSP reservoir first filled up, the reality turned out to be far removed from this vision enshrined in Plan A. One can witness buildings to house computers along the canals but in a dilapidated state; we find neither computers nor computerised water control. Over 1100 WUAs have been registered; but these are all paper WUAs which play none of the roles envisaged for them (Talati and Shah, 2004). Nowhere in the SSP command—except a handful of villages with active NGO intermediation—have farmers agreed to provide land or mobilise other resources for constructing the surface water distribution system. Indeed, in many places, the SSNNL is facing stiff resistance from farmers in acquiring land for completing even higher level distributaries and minors. The main system is all ready; yet, five years after the water was first released from the dam, Gujarat utilised only 7 per cent of the water available in the SSP (RTI India)[1]; and irrigates only around 6 per cent of its design command by gravity flow. Ironically, farmers—SSP's chief beneficiaries—themselves refuse to cooperate in creating the distribution system. Most recently, the

1. http://www.rtiindia.org/forum/4391-gujarat-used-only-7-narmada-dam-water.html

Central Government has promised funds under Accelerated Irrigation Benefits Programme (AIBP) to compensate farmers at prevailing market prices for the land needed for the distribution system. It is still doubtful if farmers will part with their land.

Plan B is in operation anyway: The absence of distribution channels, however, has not kept farmers from irrigating wherever the SSP water reached. Instead of providing land for surface channels, farmers in the SSP command have invested in thousands of diesel pumps and hundreds of thousands of metres of flexible rubber pipes to siphon or lift water from the SSP system and transport it—sometimes up to 2 km—to irrigate their fields. If 72,000 ha were irrigated by gravity flow in 2008, over 215,000 were irrigated by farmers lifting/siphoning water and transporting it by private pipelines (*Times of India*, 2009b). Pump irrigation service markets are booming. A 2008 study estimated that over 70,000 diesel pumps were purchased by farmers to pump water from the Narmada main and branch canals (Singhal and Patwari, 2008). Initially, the SSNNL watched this development with concern and even tried to use force to remove the diesel pumps mounted on canal banks (*Indian Express*, 2009). But the SSNNL was facing a dilemma: on the one hand, it was not able to achieve the irrigation targets for SSP; on the other hand, even as water was flowing away to the sea, it was preventing farmers from lifting SSP water for irrigation. More recently, therefore, SSNNL has begun to turn a blind eye to private lifting and siphoning of Narmada canal water.

Private lifting and piped distribution is rampant in all of Gujarat's canal irrigation systems. Now it is emerging as the dominant irrigation model in the SSP too. In the coming few years, we should expect a manifold expansion of this unruly system of irrigation rather than orderly gravity flow irrigation that irrigation planners and managers would like to see. This apparent chaos may be better than having no irrigation at all; but it is arguably possible to do much better by formalising piped distribution and bringing it within the SSP's regulatory framework. In view of the uncertainty about government policy regarding private pumps, as also about the water availability in canals close to them, farmers are unlikely to invest to maximise full irrigation potential. Farmers make private investments only where there is assured water supply. This is the case on main and branch canals; that is where most private pumps are installed. As a result, under the prevailing regime, the area actually irrigated by pumps and pipes may well remain confined to a 0.5-1 km strip along the main, branch and distributary canals that offer water assurance. Private investments on minors would come about only when the SSP can assure water in them when farmers need it, even if on reliable rotations.

Figures 2, 3, 4 and 5 provide an artist's imagery of the logic of the IWMI proposal. Figure 2 shows the original vision of the SSP command in the 1970s in which water was to flow into the VSA through a *pucca* (lined) sub-minor taking off from the Minor and taking water to fields through a network of field channels. Figure 3 shows the situation in the early 2000s when water began flowing in the main system but the command area remained dry for the want of a distribution system below the minor. Figure 4 shows the spontaneous and unruly development of pump-and-pipe irrigation all over the main system by farmers lifting water at will. Figure 5 illustrates what IWMI proposes: ban lifting of water from the main system; ensure weekly or fortnightly rotation of water supply in minors which may be treated as storages; and encourage farmers, co-operatives, service providers to invest in a buried pipe distribution system to irrigate the VSA.

Piped distribution from canals is a widely used technology in India and abroad: SSP planners claim that piped distribution of canal water

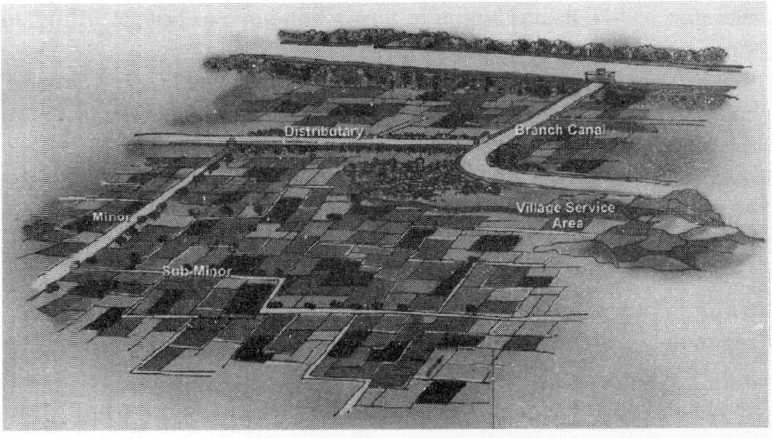

Figure 2: SSP Plan A: Water Reaches the Fields through Sub-minor and Open Channels

Figure 3: State of the Command Area Soon After Commissioning of the Reservoir and the Main System

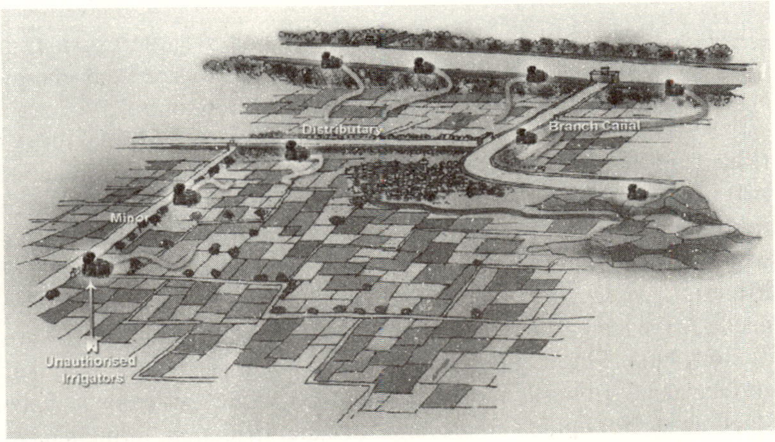

Figure 4: Chaotic Development of Command Area by Illegal Lifting of water along the Main System

is advised only in saline areas or areas prone to waterlogging. The fact is that canal irrigation systems are getting retrofitted with piped transport and distribution in many advanced countries for water and energy savings, besides other advantages (see Shah, 2009). Several irrigation systems in the Southern US, Canada, and Spain are converting water transport from surface canals to rubber-gasket jointed reinforced

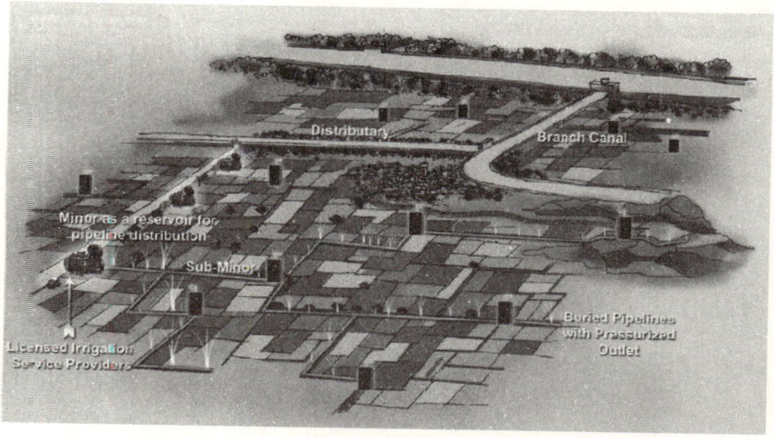

Figure 5: The IWMI Proposal: Ban Lifting from the Main System, Encourage Investment in Piped Distribution from Minor

concrete pipes (BECC, 2003)[2]. The 20,000 ha irrigation system in Hidalgo County, Texas, has proven viable, with water and energy savings, reduced health risks, and improved vector control.

Spain has similarly retrofitted its 1,000-year-old Mula system (Plusquellec, 2002)[3], and so has Alberta in Canada (Government of Alberta, 2003)[4] with buried pipes in place of open canals. The motive for pipelining in these countries is to improve irrigation service, save water from evaporation and seepage, and provide pressurised irrigation. Many of these countries are land-abundant and have no need to save land. But in the Indian context where population pressure on farm land is high, pipelining can save much productive land that surface canal networks use up. In India, too, while governments have shied away from piped conveyance in canal irrigation, farmers have aggressively been using lifts and pipes to irrigate upland areas written off as 'non-command'. In the upper-Krishna basin in Maharashtra, some 350,000 ha (gross) are irrigated by buried pipe distribution in lift irrigation systems established by individual farmers, farmer co-

2. See, www.cocef.org/aproyectos/ExComHidalgo2003_05ing.htm.
3. See, www.fao.org/docrep/004/ac799e/ac799e03.htm
4. See www1.agric.gov.ab.ca/$department/deptdocs.nsf/all/irr7197/$FILE/irrigationinalta-part2.pdf

operatives and sugar co-operatives over several distributaries of the Krishna river (Government of India, 2005). In Madhya Pradesh, the area so irrigated is even larger (Government of India, 2005).

Land Saving: The most important benefit of pipelining in the Indian context may be saving land. SSP managers often claim that the main obstacle to land acquisition is created by powerful well-off farmers. However, the ground reality is much more complex, as researchers have discovered (Talati and Liebrand, 2003). Farmer resistance would be less if all farmers were required to contribute land in some equitable manner. However, only some farmers are required to sacrifice land and not all on equal terms. Even if a few large farmers agreed, small and marginal farmers tend to be far more reluctant to give land from their already small parcels, especially now that irrigation by lifting or siphoning has made their land more valuable than ever before. Owners of private tubewells too tend to be reluctant because they have already invested in a secure irrigation source. There is also the right-of-the-way issue. Unless farmers nearest to the outlet agree, it does not help even if most farmers in a VSA agree to sacrifice some land. And it is commonly the farmers near the outlet or the head who are most unwilling because an unlined sub-minor passing by their fields threatens their land with waterlogging and flooding (Talati and Liebrand, 2003). Finally, the need felt by farmers for surface channels is less acute than was the case in the past when neither flexible pipes nor diesel pumps were as easily and cheaply available as they are today. Failure of public irrigation systems to meet farmer expectations too plays a role. There was a time when farmers happily gave a part of their land for canal construction because it would ensure water availability. But rural Gujarat today has thousands of command area farmers whose land was acquired for canals but they never got canal water.

Even at the societal level, the use of land for transporting water makes little sense in land-starved India in today's context. Creating irrigation potential would be worthwhile if productive land used up to create an irrigation system is a small fraction—say 5 per cent to7 per cent—of the land ultimately irrigated by it. If an irrigation system uses up more land than it irrigates, there is need to ask why build it at all? This is pretty much the situation in Gujarat. A recent study by Lobo and Kumar (2009:67) showed, using government data, that between 1961 and 2004, Gujarat government acquired 1.921 million ha of land—affecting 254,044 families—for constructing 21 major and 130 medium irrigation systems to create irrigation potential of 3.06 million ha (*ibid.*). However, the Gujarat government's own data suggests that in 2004, the actual area irrigated by gravity canals from major and medium irrigation systems in the state was just 0.68 million hectare—22 per cent of what was planned (Government of Gujarat, 2009). Moreover, the land benefited by canal irrigation is less than 1/3rd of the land acquired to build the irrigation dams and canal networks. It is ironic to remove 3 ha of land from productive uses to make 1 hectare more productive. It is not surprising that farmers are unwilling to part with their land for canals.

Cost effectiveness: A strong argument against pipelines is that they are costlier to lay than digging canals, especially earthen ones. But most of such back-of-the-envelope calculus of cost-effectiveness does not factor in the market value of land that canals use up and pipelines save because, in the past, the government never paid the market value of the land acquired. In the SSP, if all the 66,000 km of branch, distributaries, minors and sub-minors were to be laid, the market value of the land needed would make pipelining appear cheap. According to Lobo and Kumar (2009:72), Gujarat government has already acquired 425,000 ha of land for the SSP dam, main and branch canals. The government now claims that only 15,000 ha more are needed to build over 50,000 km of distribution network. However, this seems to be a serious underestimate since under most branches, the development of distributaries and minor canals too are held up for the lack of land. A more realistic estimate is that the completion of the entire network will likely need around 100,000 ha to cover the design command. At a market price of INR[5] 2 million/hectare, this land would be worth INR 200 billion (US $ 4.4 billion at US $ = INR 45). The pipelining project would save most of this land by replacing surface canals by underground pipelines for water transport. The investment required in pipelining would, in all likelihood, be less than the investment needed in constructing canals below the outlets *plus* the market value of land needed for open canals. If pipelines were not more attractive compared to surface channels, we would find most Gujarati farmers using surface channels for conveying water from their wells to their fields. However, today, in large parts of rural Gujarat, most farmers use buried pipeline networks to convey water from wells to their fields; even poor tribal

5. INR = Indian Rupees. Throughout this paper, US $ = INR 45

farmers use overland pipes in place of earthen channels to convey water. In central and north Gujarat, tubewell owners invest in elaborate buried pipeline networks to irrigate their own lands as well as the land of scores of other farmers (Shah and Bhattacharya, 1993). In Navli, a small village in Anand district, Shah (1993) estimated that private farmers had invested in 65 kilometres of buried pipeline networks to irrigate 1200 ha of land by tubewells. In Gandhinagar, Mehsana, Patan, Kheda, Anand and Vadodara districts, private investments of this order in pipelines are common in most villages. This would not have happened had the cost-benefits of piped irrigation been as adverse as it is made out to be.

Water Saving Impact: In semi-arid climates with high wind speeds, which is the case in much of Gujarat, surface storages and canal networks suffer high non-beneficial evaporation from open surfaces. Annual pan evaporation rates in Gujarat as a whole are well over 200 cm/year but in Kachchh and North Gujarat, these can be over 300 cm/year (Krishiworld, 2010)[6]. This implies that a body of standing water exposed to sun loses at least 2 metres/year just to evaporation. A surface irrigation system in western India is thus a vast evaporation pan. SSP's 37,000 ha reservoir and some 20,000 ha of canal network—much of which is exposed to sun throughout the year because the canals provide drinking water year-round—is likely lose around 0.5-0.6 BCM of water annually to non-beneficial evaporation. An equivalent amount of groundwater provides supplemental irrigation to over 100,000 ha of land. Indeed, minimising non-beneficial evaporation of water in storage and transport can be a major pathway to improved water productivity at basin level. As a fraction of the water they carry, the minors, sub-minors and field channels lose most to evaporation because their ratio of surface area to depth is much higher than it is for the reservoir, and the main and branch canals. In so far as the pipelines will replace the most-evaporating tertiary canal network, it can potentially save annually anywhere up to 0.5 BCM of water that the SSP would lose to evaporation under surface distribution system. There would be a similar volume of water removed through seepage; however, much of this ends up as highly valuable groundwater recharge which Gujarat badly needs, and therefore, cannot be viewed as a loss.

6. http://www.krishiworld.com/html/water_resources2.html

Water Productivity Impact: Reducing evaporation from open surfaces can be a big boost to basin water productivity in the state. However, pipelining can also contribute greatly to the irrigation productivity of water applied. Surface irrigation systems in Gujarat, as indeed in India as a whole are highly inefficient when compared to groundwater irrigation by private wells and tubewells. Table 1 shows data for Gujarat (Government of Gujarat, 2009) to illustrate the point.

For a net hectare *actually* irrigated by canals, Gujarat maintains reservoir storage of nearly 35,000 m^3 whereas for a hectare irrigated by wells and tubewells, the state uses all of 3300 m^3 of groundwater storage. It is sometimes argued that reservoir storage is used not only for irrigation but also for drinking water and industrial needs. The counterargument is that this is even truer of groundwater. Farmers tend to use groundwater more carefully because rationed electricity supply imposes a quantity constraint and high cost of diesel imposes a surrogate price constraint. For these same reasons, piped distribution of SSP water will encourage farmers to use canal water as efficiently as groundwater.

Table 1: Surface and Ground Water Development and Use in Gujarat

Sl. No.	Details	Surface water	Ground water	Total
1.	Total rainfall precipitation in a normal year (MCM)			1,96,000
2.	Water resources (MCM/year)	37,400	15,800	53,200
3.	Water resources available for utilisation or utilised (MCM/year)	27,485	11,500	33,695
4.	percentage of domestic water requirements met	18	82	100
5.	Percentage of industrial water requirements met from	35	65	100
6.	Storage available for irrigation (80%) (assumed) (MCM)	21,988	9,200	26,956
7.	Net area irrigated (million ha)	0.63	2.76	3.39
8.	Storage volume/ha (m^3)	34,900	3,300	7,952

* Surface water storage in Major, Medium and Minor irrigation projects plus recharge structures (check dams, percolation tanks, etc).

Conjunctive Use Impact: One big opportunity that canal irrigation offers is of conjunctive management of ground and surface water to avoid problems arising from excessive use of either surface or groundwater in irrigation. But this requires that available surface water is spread over as large an area as possible. As we saw earlier, the water in Gujarat's surface storages reaches and wets only a fifth of the area it is designed to irrigate. As a result, large swathes of Gujarat's agricultural lands are exclusively dependent on groundwater irrigation without any benefit of canal recharge, resulting in severe groundwater depletion. Spreading canal water over as large an area as possible can alleviate Gujarat's groundwater depletion problem. A pipeline-based distribution system can spread a given volume of water in surface storages over a larger area through pumping. This could in theory bring much of Gujarat's groundwater irrigated farm lands under a conjunctive management regime. If the 27.5 BCM of Gujarat's surface storage (including the SSP) were to be spread over 3.5 million ha of the state's net irrigated area through a total pipeline-based system (i.e. without significant non-beneficial evaporation) at a rate of 7,800 m^3/ha, it would nearly eliminate the need for pumping groundwater to that extent; and moreover, a portion of that surface water would end up recharging the aquifers.

Energy Impact: Against all these advantages of pipelining, its chief disadvantage is its energy requirement. As C.C. Patel has argued, in the absence of hydraulic head in flat lands, it needs energy to push water to counter the friction loss in pipes, even if pipelines are laid such that they use the head that the distribution canals were to use for gravity flow. Hemant Shah (2009) has suggested that pipes may have to transport water on average 10-15 km from branch canals and that an additional 1500 MW of energy will be needed to convey 9.8 BCM of water to 1.8 million ha. This is a serious overestimate; however, even if we take it as correct, we need to compare this with the 6300 MW of electricity farmers are currently using to operate Gujarat's 800,000 tubewells[7] (Government of Gujarat, 2009).

For a gigantic project such as the SSP, the real issue is whether a pipelined SSP can have a favourable influence on Gujarat's overall irrigation energy balance sheet. Gujarat's 800,000 electrified tubewells today use around 12 billion kWh of electricity (valued at some INR 60 billion or US $1.33 billion/year at generation stations)—to lift some 7 to 8 BCM of groundwater. Another 480,000 diesel pump

7. Assuming an average load of 10.6 hp/tubewell (Government of Gujarat, 2009).

owners use around 200 million litres of diesel/year (valued at INR 7.2 billion or US $ 160 million) to lift 1-1.5 BCM of shallow groundwater and some surface water. Gujarat's groundwater-based agriculture is thus extremely energy intensive anyway. The only way to curtail this energy intensity is to spread available surface water over large areas of North Gujarat, Saurashtra and Kachchh which would produce a double-effect: (a) availability of surface water would reduce need to pump groundwater; and (b) surface irrigation return flows would enhance groundwater recharge and thereby reduce the pumping head in tubewells. This can be done only by pipelining the distribution system in not just the SSP but also other canal systems. An indication of energy savings that such a strategy can produce is that all the power needed by a pipelined SSP would be recovered if it could put only around 25,000 deep tubewells in North Gujarat and Kachchh out of business.

Finally, a major new objective SSNNL is working for is promoting micro-irrigation in the SSP command to maximise water use efficiency. If the command area farmers were to use micro-irrigation, they would have to use pumps to create the pressure needed to operate drips or sprinklers even if water were delivered to their fields by gravity. Micro-irrigation can be integrated far more easily in a piped distribution system than in a gravity flow distribution system.

Public Private Partnership: By far the most important advantage of pipelining water distribution in SSP is that given encouragement and supportive policy environment, farmers and their co-operatives will take over the entire responsibility of investing in pipeline projects and managing them. The best proof that this will happen is offered by the experience of irrigation development in Kolhapur, Solapur, Sangli, Satara, Nanded, Nasik and Ahmednagar districts in the upper Krishna basin in Maharashtra. We describe this experience in some detail in the paragraphs below.

In 1976, the Bachhawat Award allocated 560 TMC[8] of water to Maharashtra which the state had to develop by 2000. Maharashtra was not in a position to build reservoirs and canal networks needed to use this water; by 1996, it had constructed only 385 TMC of storage and little by way of canal network in the Krishna basin. Therefore,

8. Thousand Million Cubic feet.

the government first began allowing farmers to lift water from Krishna and its tributaries. But this only encouraged small-scale private lift schemes most of which could not convey water beyond a distance of 1-1.5 km. In 1972, only 200 private and co-operative lift schemes were operating in Maharashtra. As pressure to utilise the water mounted, the government adopted a more proactive posture towards lift irrigation schemes. It introduced a capital cost subsidy for irrigation co-operatives and also facilitated bank finance from nationalised and co-operative banks. Most importantly, the Irrigation Department (ID) constructed a series of Kolhapur Type (KT) weirs across many tributaries of Krishna to use them as storages for lift irrigation schemes. Each such scheme has to be approved by the ID, after which it qualifies for an electricity connection and bank finance. Each scheme also has to pay irrigation fees to the ID for the actual area irrigated and pay the electricity charges to the State Electricity Board at prevailing rates for agricultural use. Between December and June each year, the ID implements a fortnightly schedule of water releases to fill up the dykes, starting with the last dyke first. This ensures that lift schemes will have access to reliable water supply during the irrigation season.

A good example of the kind of partnership between the ID and irrigation co-operatives that Maharashtra's policies have spontaneously promoted is seen in the Radhanagari project that serves 91 villages in Kolhapur district. The dam never had any canals; water is released from the dam into Bhogavati River on which the ID has constructed a series of KT weirs. The ID has three roles: (a) approve proposals for new schemes; (b) release water into Bhogavati river every 15 days to fill up all the KT weirs; and (c) collect irrigation fees from all lift schemes based on crops grown and area irrigated. Water lifting, conveyance and distribution are all done by some 500 'Irrigation Service Providers (ISPs)' in private and co-operative sectors.

This scheme is studied by Choudhury and Kher (2006), Padhiari (2005) and Chandra and Sudhir (2010). These found that Radhanagari's performance over the past two decades has been very good compared to surface irrigation systems anywhere in India. Against a design command of 26,560 ha, the average area irrigated by ISPs during 2001-2006 was 30,341 ha/year. However, the ID managed to collect only 58 per cent of the irrigation charges that fell due. Against the annual O&M cost of INR7.9 million (UD $ 0.18 million), irrigation charges collected in 2005-06 were INR 17.9 million (US $ 0.4 million). In terms of the area irrigated as well as irrigation charges recovered, tail-end areas were found no worse off compared to head; the practice of filling up KT weirs last to first seems to address the head-tail inequity. An informal survey suggested that the number of irrigations the project provides is 80 to 90 per cent of the number needed and that over 80 per cent of the farmers interviewed were happy with irrigation provided by the ISPs (Choudhury and Kher, 2006). In terms of offering irrigation-on-demand, Radhanagari comes close to tubewell irrigation.

Choudhury and Kher (2006) interviewed 8 private and 9 co-operative ISPs that irrigate a little over 1,000 ha in Radhanagari project. These ISPs had altogether invested nearly INR 220 million (US $ 4.9 million) in systems that included 2,280 hp of pumps and 41 km of buried pipe network, and employed 92 staff to manage water. Typically, every system has a rising main—sometimes, multi-stage—to a chamber from where water is conveyed by buried pipes to fields. These ISPs thus invested INR 220,000/ha in the system, use 2.3 hp/ha of power load, employ a water manager for every 12 ha irrigated and collect an irrigation charge that is high enough to pay off their debt, pay electricity charges to the Electricity Board, irrigation charges to the ID and salary to employees. After these, they still save enough for prompt repair and maintenance.

Radhanagari may appear an exception but is not. According to Government of India's Minor Irrigation Census III, in 2000-01, Maharashtra had some 100,000 such schemes in operation for lifting and piped distribution of surface water mostly in the Upper Krishna basin (Government of India, 2005). Over 20,000 of these were owned and operated by farmer groups and co-operatives. These lifted water from rivers and streams and transported it mostly by buried pipelines to up to 30 km from source. Remarkably, none of these was operated by a government agency. Over 90 per cent of Maharashra's lift schemes were constructed by farmers from their own funds and bank finance, with the present value of aggregate investment of around INR 50 billion (US $ 1.1 billion). Over 90 per cent of the schemes used electric pumps to lift water and 70 per cent had buried pipeline network for water distribution. Total horse power of pumps installed in these schemes was around 590,000, equivalent to 440 MW, even though all the schemes involved a sizeable lift ranging from 20 metres to 185 metres. These irrigated a gross area of some 350,000 ha (including sugar cane area of over 100,000 ha). Maharashtra's lift irrigation schemes likely employed over 100,000 workers as *pankyas* (water managers), if we count the fact that the 80,000 families operating private lift schemes had at least one family member each devoted full time to work on the scheme operation.

Discussion

A discussion paper outlining the arguments above (which we may call the 'IWMI Proposal'[9] for convenience) was shared with technocrats and political leaders of the Government of Gujarat. Thereupon, the Chief Minister constituted an "Expert Group for Strategy for the Accelerated Development of Sardar Sarovar Project Command Area". An Expert Group with a broad cross section of expertise in fields such as agriculture, groundwater, economics, planning, irrigation engineering would have been required to do justice to the IWMI Proposal. However, the seven members of the Expert Group, including the chairman, were all highly experienced irrigation engineers best known for their knowledge and experience in design, construction and management of government canals[10]. The Expert Group met three times. In the intervening period, sub-groups made brief field visits to several government-operated irrigation pipeline schemes. The results of these visits were reflected in the minutes of the meetings. Based on this work, the Expert Group

9. After International Water Management Institute which led the research.
10. The Expert Group was chaired by B.N. Navalawala, former Secretary, Water Resources, Government of India and presently Water Adviser to Gujarat's chief minister, a well-known water resource planner who played a key role not only in planning the SSP but also in shaping India's irrigation policy during the 1990s. Other members included A.S. Kapoor, N.B. Desai, R.G. Kulkarni., B.J. Parmar, K.S. Srinivas. All except K.S. Srinivas were canal irrigation experts. The Expert Group also invited inputs from distinguished irrigation engineers like M/s C.C. Patel, V.B. Patel, M.U. Purohit, M.R. Goswami, all very well known for their experience and expertise in canal irrigation.

rejected the IWMI Proposal by recommending that "in SSP (the) command area the conventional open channel should be adopted for construction of sub-minors" and buried pipelines should be used only when open channel is not feasible. Two more recommendations made had to do with measures to expedite land acquisition and construction of the open channel distribution system below the outlet.

The rejection of the 'IWMI Proposal' was based on the experience of Vatrak, Karjan and Ukai Right Bank systems, three large irrigation projects where underground pipelines were used under gravity flow in portions of the command area. Poor performance here was attributed by the Expert Group to poor operation and maintenance, siltation, insufficient ground slope and available head, and farmer vandalism. However, the Expert Group also considered evidence from the Narmada Canal System in neighbouring Rajasthan state, where water is delivered by canals into *diggis* (ponds) from which it is pumped and carried through buried pipes for sprinkler irrigation. The pumps and buried pipelines are installed by the government but will soon be handed over to WUAs for management. Farmers buy sprinkler sets with government subsidy. The original motive of the scheme was to reduce canal irrigation supply to a third of the originally planned water allowance and encourage farmers to use it conjunctively with groundwater, an aim that is valid for the entire command area of SSP since the system is planned for a water allowance of just 21 in. per hectare during the irrigation season. Visits to the Rajasthan projects by the Expert Committee members showed that these worked 'very well', and that the farmers paid the full electricity cost. "The team members were entirely satisfied with innovative system put into operation by the Rajasthan Government and congratulated the engineers, farmers and stakeholders for boldly implementing the scheme" (Government of Gujarat, 2010:129). Based on this field visit, one of the Expert Group members developed a Public Private Partnership model for water distribution which in essence is very close to the IWMI Proposal: canal managers deliver regular water supply at the outlet on a roster of weekly or fortnightly *warabandi* (rotation); the private ISP ensures regular volumetric water supply to each field from *diggis* and keeps the pipeline system well-maintained; farmers or their groups pay for energy and water to the ISP (Government of Gujarat, 2010:149-151).

Table 2: Expert Group's Critique and our Responses on 10 Reasons in Support of Pipelines

Sl. No.	*IWMI Argument*	*Expert Committee Response*	*IWMI Response*
1.	Plan A is not working because farmers are reluctant to part with the land	Not true; it is working in Phase I; in other five phases, distribution network construction is delayed.	The plain fact is that seven years after the dam and the main and branch canals were ready, less than 10% of the design command of SSP is irrigated.
2.	Plan B is in operation anyway with tens of thousands of farmers lifting water and conveying it by pipelines	Only in certain places and in some area where water is not flowing properly up to the tail end.	Our surveys found that private lifting and piped distribution of water in SSP is rampant. Numerous newspaper reports too confirm this. If this were no so, the Government would not have deployed mounted police on Narmada canal to check such 'unauthorised' lifting.
3.	Piped distribution is a widely used technology in India and abroad	Pipelines are used where no gravity flow irrigation is available. Pipelines are useful for tubewells but not relevant for a vast project like SSP.	Thousands of kilometres of pipelines are used by farmers on canal systems as well as or tubewells in Gujarat's canal command areas. Diesel pump dealers affirm that some 70,00 pumps were purchased in recent years for use on the SS system.
4.	Piped distribution can save land	The land required for creating surface distribution system is insignificant.	If so, why is such a large, prestigious project delayed due to inability to acquire land for building sub-minors?
5.	Piped distribution can be relatively cost effective if the land cost is factored into surface canals	Pipelines too require land under the right-of-use act; farmers will demand compensation for crops lost during operation and maintenance.	Thousands of km of private irrigation pipelines neither pay for right-of-use nor crop compensation. These are demanded only when government installs pipelines; doing this should be left to private ISPs.
6.	Piped water distribution can save evaporation losses at sub-minors where the surface area to depth ratio is the highest.	SSP water is to be provided only during ambient temperatures that do not permit high evaporation.	Canal network is carrying water during past five summers. More evaporation losses from a sub-minor network of over 50,000 km would mean a shallow exposed water body of 10,000 ha. Replacing them with pipes will save on evaporation.

Sl. No.	IWMI Argument	Expert Committee Response	IWMI Response
7.	Piped distribution can raise water productivity by spreading canal water over a larger area	Groundwater distributed through pipelines has high productivity because farmers have to 'pay through their nose' for it.	Piped distribution of SSP water too will attract a volumetric charge as is common with all private service providers; it will therefore be more productive than gravity flow irrigation which will not be charged on volumetric basis.
8.	Pipes can promote conjunctive use of ground and surface water over large areas; surface canal command shrinks over time.	This is not true; surface channels will take water to every field in the design command; moreover, groundwater is used only in small pockets where available in abundance.	No canal irrigation project in Gujarat today is able to irrigate more than 40-50% of its design command; groundwater irrigation is pervasive throughout the state and can benefit from conjunctive management with surface water.
9.	Piped distribution is energy intensive but can also save energy by reducing groundwater pumping and promoting conjunctive use.	Gravity is the cheapest form of irrigation and a huge gift from God.	The Report ignores the argument that spreading surface water over large areas in North Gujarat can raise groundwater levels and reduce energy costs of pumping. It also ignores the potential to link tubewell pumps to SSP minors for dual source irrigation.
10.	Piped distribution can promote public-private partnerships in ways that gravity flow has not done.	The argument is based on 'some isolated cases ...quoted to impress'. Surface canals are the 'easiest, cheapest and most effective' mode to transport water.	The Maharashtra private pipeline schemes are in tens of thousands; in North Gujarat, pipeline schemes operated by tubewell cooperatives too are in tens of thousands (Shah and Bhattacharyya, 1993). These have evolved their own institutional models for sustainable management of piped water distribution.

In our assessment, then the Expert Group's report did not satisfactorily address any of the 10 arguments in the IWMI proposal (see Table 2 above). Moreover, its rejection of the IWMI Proposal does not follow from the evidence its members gathered. All the cases of poor performance of buried pipelines it found were under government management and 17 depended on gravity flow. It did not consider the evidence within Gujarat of some half a million kilometres of private, farmer-managed pipeline networks which have, for decades, overcome all the problems of operation and maintenance, siltation and clogging, obtaining right-of-the-way from other farmers, and also overcame problems of farmer vandalism which make government-managed pipeline systems perform poorly. The Expert Group discarded the experience of piped distribution in Narmada canal in Rajasthan by casting doubt about the future operation and management issues once the pumps and pipelines are handed over to WUAs. In our thinking, such doubts are bound to arise when government installs pumps and constructs pipeline networks; if this is done by farmers themselves, they will ensure efficient operation and maintenance as they have shown for decades (Shah, 1993). The Expert Group suggests that pipelines are energy-intensive. We agree, but the Expert Group totally ignored the aggregate energy saving from conjunctive use of rationed canal water supplies and deep groundwater. Above all, the Expert Group's report remains blind to the profoundly changing nature of Gujarat's irrigation economy and the deteriorating performance of canal irrigation systems. It also ignores the contribution that such innovative hybrid systems as they found in Rajasthan can make to reversing the trend. Indeed the principal ground on which the IWMI Proposal is rejected is that "The entire Narmada Project has been conceived on the premise that water would be delivered through flow irrigation only." It mattered little to the Expert Group that Gujarat's irrigation economy has vastly changed from the time when SSP was conceived during the 1970s. It also mattered little that the Rajasthan pilot, which departed from 'the premise' is, in their own assessment, doing 'very well'.

Summary and Conclusion

In this chapter, we have argued that there are at least 10 strong reasons why Gujarat should seriously consider the IWMI proposal for pipelining the water distribution system below the minors in the SSP command. The IWMI Proposal, in brief, argues that: (a) SSNNL should abandon the construction of sub-minors below the outlet; (b) it should incentivise private or cooperative ISPs to invest in water distribution systems based on buried pipelines; (c) SSNNL should focus its management effort on delivering bulk water at the minor level on a volumetric basis in a strictly enforced, predictable rotational schedule rather than delivering water on each field. Doing this will develop SSP command area faster and better. Moreover, besides saving land, water, energy and public funds, pipelining opens up huge possibilities for public-private partnerships and farmer participation in irrigation management in ways that surface canals have failed to provide. For farmers and co-operatives to become ISPs, four conditions are necessary: (a) potential for economic value creation through irrigation service provision; (b) assurance of reliable water supply; (c) water supply on a schedule to match the irrigation demand pattern; (d) freedom to market or share 'irrigation service' as a private good. These conditions prevail under piped water delivery and not under gravity flow through open channels.

The proposal that Government of Gujarat's Expert Committee rejected outright, involved the SSNNL constructing and managing a pipeline system for gravity flow distribution—something inimical to the IWMI Proposal. Pipeline-based lift irrigation systems need intensive management; as a result, government-managed piped lift irrigation schemes have met with resounding failure in Gujarat and elsewhere in India. If the SSNNL were to install and manage piped distribution, chances are high the project would fail. Instead, it should emulate the Maharashtra example to invite farmers and their co-operatives to invest in and manage piped distribution systems. To promote farmer investments in piped distribution in a planned and systematic manner, SSNNL should: (a) recognise and incentivise lifting of water from SSP minors and its piped distribution; (b) ban water lifting from the entire system except at the minor level; (c) register all ISPs, private, co-operative and others, and bring them under formal regulatory ambit; (d) make firm commitments of weekly or fortnightly water deliveries in each minor along a strict schedule, *a la* farm power scheduling under *Jyotigram*[11]; (e) encourage existing tubewell owners to use their electricity connections to lift water from SSP minors; (f) persuade the state electricity utility to provide new electricity connections to approved

11. Jyotigram is an innovative programme implemented by the government of Gujarat to ration electricity supply to tubewell irrigators to curtail electricity subsidy burden and groundwater overdraft without adversely affecting groundwater irrigated agriculture. Under this programme, the electricity utility rations power supply to tubewells to 8 hours/day; however, power supply is provided on a pre-announced roster, with minimal interruptions and at 440 voltage. Improved quality and reliability of power supply has helped tubewell owners accept the stringent rationing on supply (Shah and Verma 2008; Shah *et al.*, 2008)

piped distribution schemes planned by farmers, co-operatives and producer companies; (g) invite NABARD[12] and other institutional financial agencies to finance farmer cooperatives for their investments in pipelines; (h) offer a subsidy (we suggest 25%) on capital costs of approved projects in ways that minimise perverse incentives; (j) register each pipeline with the SSNNL and require it to pay irrigation fee—preferably volumetric—for all the land irrigated with SSP water; (k) revise the traditional idea of 'irrigation command' to include all areas where farmers are willing to invest in piped distribution system and pay volumetric water charges.

Fortunately, SSP is designed as a network of Constant Volume Canals (CVC) up to minors; as a result, as water gets withdrawn from any minor, upstream gates automatically open to replenish the water released. Every minor, distributary, branch and main canal of the SSP is in a dynamic hydraulic equilibrium with the reservoir. This implies that SSP minors can be used as excellent storage canals, needing little modification to support piped distribution. If undertaken with farmer investment and management thus, pipelining can: (a) achieve SSP's potential quickly, at little cost to the government; (b) promote large scale ISPs that would invest in long distance pipeline networks; (c) create some 150-200 thousand rural jobs in water distribution; (c) spread SSP water over a large area; (d) enable conjunctive management of ground and surface water; (e) improve Gujarat's overall farm power budget in net terms; and (f) establish a new model of irrigation management partnership between farmers and the government. This would also be the best way to encourage and facilitate the spread of pressurised micro-irrigation in SSP command. Above all else, this can achieve SSP goals without additional investments of public funds. All that SSNNL needs to do is to make a credible commitment of reliable—even if rotational—water availability in every minor.

Acknowledgements: This study was supported by IWMI-Tata Water Policy Programme funded by IWMI and Sir Ratan Tata Trust, Mumbai. Authors gratefully acknowledge useful comments from Dennis Wilchens, David Molden, Avinash Kishore, Sam Johnson and Martin Burton.

12. National Bank for Agriculture and Rural Development which provides institutional finance to such undertakings.

28

Maximising the Social and Economic Returns from Sardar Sarovar Project: Thinking beyond the Convention

M. Dinesh Kumar and *Nitin Bassi*

Changing Socio-ecology of SSP Command: Imperatives for Future Management

Development of command areas and catchments is vital to realising the full potential and sustaining the performance of large irrigation systems. The standard approach to command area development in India is based on the premise that surface irrigation would lead to excessive seepage and return flows from canals would lead to waterlogging and salinity problems in the absence of sufficient control on water delivery and farmers shifting to water-intensive crops such as paddy and sugarcane. An associated concern is that water distribution would be highly inequitable across the hydraulic system, with more water being made available to the farmers at the head reaches and very little or no water reaching the tail-end farmers. As a consequence, suggestions for CAD schemes focused on lining of canals, building of drainage networks, promotion of conjunctive use of surface and groundwater, and creation of farmer organisations for equitable distribution of water among the farmers in the command area, along with many other irrigation management functions.

Such views emanating from the discussion on efficient management of irrigation were rational at a time when well irrigation was not in vogue, groundwater was under-developed, pump irrigation was too expensive owing to scarcity of electricity in rural areas, and the poor advent of energised pump sets, water was not as scarce a resource as it is today in many regions, and infrastructure and technologies to access water were a major issue.

Today the problems are different. Groundwater resources are fully developed in the entire state, which includes SSP command. Groundwater over-exploitation has become one of the greatest environmental problems facing the state, with depletion and quality deterioration problems. They are more severe in central and north Gujarat (Kumar, 2007), which are likely to get the least benefit from SSP in terms of area irrigated (Raj, 1991). The agriculture sector has become one of the largest consumers of electricity, which it uses for pumping groundwater, and the state's revenue losses in the form of electricity subsidy for pumping groundwater amounts to nearly Rs. 40 billion annually, which increases every year with decline in groundwater levels across the state. A major chunk of this goes to central and north Gujarat, where water tables are deep. Under such situations, as pointed out by Allen *et al.* (1998) and Chakravorty and Umetsu (2003), seepage from canals cannot be treated as a waste or a negative externality. Instead, it becomes a valuable resource, which can be re-used through wells.

On the other hand, due to excessive diversion of water, the rivers of north and central Gujarat are experiencing one of the severest environmental water stresses. Groundwater quality deterioration affects thousands of villages in north and central Gujarat, which are exposed to poor quality groundwater for drinking, with a myriad of water-related health problems such as kidney stone and fluorosis (IRMA/UNICEF, 2001).

While on-farm water management and equitable water distribution would still be great challenges to be addressed in irrigation management, there are far greater challenges of maximising the economic, social and environmental benefits from the project in view of the positive externalities that large water resource projects like SSP can inflict on society. As Shah and Kumar (2008) note, large water systems in developing economies like India can and do produce multiple benefits, in addition to irrigation, hydropower and water supplies. While some of them are unintended, some are unforeseen.

A few examples that can be cited are improved groundwater recharge, rise in groundwater levels, enhanced cereal production, clean energy, increased inland fish production due to year-round water availability, drought-proofing, and increased biomass production due to energy plantations (Shah and Kumar, 2008). Given the fact the water and energy environment in Gujarat had drastically changed since the time SSP was planned, the economic value of these benefits can outweigh the direct economic benefits of the project. But, there is little appreciation of these externalities (Kay *et al.*, 1997; Shah and Kumar, 2008).

SSP can make a positive contribution to improving groundwater recharge in over-exploited areas, sustaining well irrigation and reducing well failures (Ranade and Kumar, 2004; Shah and Kumar, 2008); reduction in cost of pumping groundwater (Shah and Kumar, 2008; Vyas, 2001); reducing carbon emissions through generation of clean energy (Joshi and Kapadia, 2010); improvement in groundwater quality and the resultant improvement in water supply situation in villages and urban areas with positive health benefits; improving the health of riverine ecosystem by the release of water into the rivers of north and central Gujarat (Shah and Kumar, 2008; Desai and Joshi, 2008); and lowering of cereal prices (Perry, 2001a; Shah and Kumar, 2008); and gain to the state exchequer in the form of reduced electricity subsidy for farm sector.

But not all benefits can be realised uniformly across the hydraulic system. Groundwater recharge is an important benefit in an area where the resource is mined. Groundwater remediation benefits of recharge will be significant in areas where the natural quality of water is poor or moderate, but not extremely poor. The SSP command cuts across several agro-climatic sub-regions (Raj, 1991) and also several geo-hydrologic environments (Kumar, 2002). The south Gujarat part of the command receives high rainfall (1000-1200mm), has sub-humid climate and shallow aquifers, whereas the north and central Gujarat parts of the command have low to medium rainfall; semiarid to arid climate and heavily-exploited deep alluvial aquifers. Saurashtra has low to medium rainfall and shallow, hard rock aquifers. Rivers in south Gujarat are perennial, while those in central and north Gujarat and Saurashtra experience environmental water stress (IRMA/UNICEF, 2001; Kumar, 2002).

The extent to which we can realise these benefits depends on how judicious is the allocation of water from SSP across regions. It is quite clear that the usual interventions taken up under command area development (CAD) programme would not help. On the contrary, they can often be counterproductive if implemented on a blanket basis. This means that the objectives of CAD programmes and criteria for assessing them should be broadened to internalise the changing socio-ecological realities of the region. The following discussion would illustrate why the conventional approach is not going to bring about optimum benefits and a unique strategy is needed in the case of SSP, to maximise the economic, social and environmental benefits from the project both direct and indirect.

Water Scarcity in Gujarat: Various Dimensions and the Dichotomy[1]

Water problems can be analysed from various angles: from a pure physical perspective, from an economic perspective (economic scarcity—often called financial scarcity), from the perspective of demand and supply of water, or from an environmental perspective. Scarcity problems in Gujarat are analysed from the point of view of physical availability of renewable freshwater, available water supplies, and the demand for water from various sectors and the extent of diversion and use of renewable fresh water. The estimates of renewable per capita freshwater availability for the year 2001, water use for the year 1997 and water requirements and supplies for the year 2025 are drawn from the authors' own analysis for the Draft White Paper on Water in Gujarat (IRMA/UNICEF, 2001), details of which are available in the document.

Physical Scarcity of Water in Gujarat

The total renewable freshwater available in Gujarat, including the annual runoff from within Gujarat, that is allocated from the neighbouring states, and all the natural recharge to groundwater, is

1. This section draws heavily from Kumar and Singh (2001), "Market Instruments for Demand Management in the Face of Scarcity and Overuse of Water in Gujarat", *Water Policy*, 3 (5).

54,593 MCM. This gives a per capita renewable freshwater availability of 1137 m^3 per annum for the year 2001. Going by M. Falkenmark's indicator of physical water scarcity, the state is water stressed (as it is below 1700 m^3 per capita) and is likely to face occasional scarcity. But the availability of water is heavily skewed towards south and central Gujarat which has 69.5 per cent of the total renewable freshwater. Therefore, it is important to adopt a regional approach, if one needs to assess the magnitude of water scarcity from a supply-based approach. Going by regions, the per capita renewable freshwater availability in north Gujarat is only 427 m^3. Thus, the region is "absolutely water scarce" (as it is below 500 m^3 per capita) where water becomes a constraint to life. Saurashtra and Kachchh have per capita renewable freshwater availability of 787 m^3 and 875 m^3 respectively. Thus these regions are "water scarce" (below 1000 m^3 per capita) and water becomes a constraint for economic growth and social advancement.

However, it is important to remember that the estimation based on renewable water availability does not take into account the groundwater stock in regions such as north Gujarat. It is quite significant when compared to the renewable groundwater of the region. Therefore, these numbers of annual per capita renewable water availability have to be used very cautiously when one looks at the implications of such scenarios for the use of water for economic development in the region. Obviously, the static groundwater resources have played a major role in sustaining the agricultural economy in the region for many decades since its intensive use started. But today most of these static resources are mined with withdrawal of groundwater for agriculture far exceeding the renewable groundwater. Therefore, for projecting future scenarios of human development and economic growth based on water, such numbers would still be quite relevant.

Demand Driven Water Scarcity

According to the estimates prepared for the White Paper on Water in Gujarat, the total water requirement for agricultural production, domestic and industrial use and consumption by livestock in the year 1997 was 27,622 MCM. The projections showed that by the year

2025, the total requirement of water for agriculture, industry, livestock and domestic purposes would be 53,050 MCM. Here it was assumed that the prevailing patterns of water use would continue in the future, or in other words, a "business as usual" scenario was generated. Under this scenario, we found that even after harnessing all the potentially utilisable surface and ground water resources (100% of the surface water and 90 per cent of the utilisable groundwater) including water from the neighbouring States of Madhya Pradesh, Maharashtra and Rajasthan and water from Sardar Sarovar Project (11,100 MCM), the supplies would be only 41,250 MCM. This leaves a gap of 11,837 MCM. The situation would appear rather bleak, if one does not consider the allocation of water from Narmada, where the magnitude of gap will be 22,937 MCM.

Now, the per capita withdrawal of water is found to be totally unrelated to the relative water availability in a particular region. For instance, the maximum withdrawal of water for human uses was found in the case of North Gujarat (448 m^3) which has the lowest per capita renewable freshwater (447 m^3). The minimum per capita water withdrawal is in South and Central Gujarat (372 m^3) where per capita renewable water availability is highest (1932 m^3). This means that if current water use patterns continue, the most severe scarcity of water is likely to occur in the region where its physical availability is low.

Scarcity Perpetuated by Unsustainable Water Use

A study commissioned by the UN Commission on Sustainable Development (Raskin *et al.*, 1997) had defined water scarcity in terms of the total amount of annual withdrawal as a percentage of the Annual Water Resources. By this criterion, if the annual withdrawal exceeds 40 per cent of the Annual Water Resources, the region can be called water scarce.

Going by this criterion, three regions of the State, namely North Gujarat, Saurashtra and Kachchh are already water scarce and the figures are alarming. In North Gujarat, for instance, the average annual water withdrawal for all uses (estimated to be 6008 MCM) is already close to the Annual Water Resources (6105 MCM), giving clear signals of unsustainable resource use. As the runoff (which accounts for 33 per cent of the total renewable water) is not fully exploited, the additional water for withdrawal is coming from depletion of the stored groundwater. Alarming drops in groundwater levels manifest this. In Saurashtra, the average annual withdrawal of water is estimated to be 5,149.5 MCM against a total renewable fresh water availability of 9,287 MCM. In Kachchh, the average annual withdrawal is 691.7 MCM against an average annual freshwater availability of 1,275 MCM. The withdrawal is exceeding 40 per cent of the renewable fresh water in all the three regions and therefore, going by the criterion of Raskin *et al.* (1997), these three regions are water scarce.

The Sardar Sarovar Project: Challenges Facing an Economic Powerhouse

The foregoing analysis of regional water situation clearly points to the fact that SSP is among the rarest of the rare hydraulic infrastructure projects in the world that have the capacity to level off the regional imbalances in water availability and demand in a large geographical area of 180,000 sq. m. It can change the agrarian economies of all the four regions of Gujarat, namely south Gujarat, north and central Gujarat, Saurashtra and Kachchh. Though the project has already become the major driver of Gujarat's agricultural growth (Kumar *et al.*, 2010), it faces some major challenges.

SSNNL is not charging for irrigation water on volumetric basis in the area where the commands are receiving water. The farmers, therefore, have strong incentive to go for crops that are highly water-intensive such as banana, even in the naturally water-scarce regions of north and central Gujarat, and also use water excessively. Such a trend has been seen in well irrigation in Gujarat (Kumar, 2005), UP and south Bihar (Kumar, 2009) under flat rate pricing of electricity wherein farmers are confronted with zero marginal cost of using electricity and therefore groundwater. SSNNL needs to create infrastructure and put institutional mechanisms in place to make measurement and volumetric pricing of irrigation water a reality. Earlier analysis had shown that farmers in south Gujarat who grow highly water intensive banana and sugarcane incur much lower cost per unit volume of irrigation water as compared to those who grow low-water consuming crops, due to the inefficient (crop-area based) pricing in which irrigation of water consuming crops attract higher charge per unit area (Kumar and Singh, 2001).

Direct lifting of water from canals by farmers who are both inside and outside the designated command area is emerging as a major challenge for SSNNL officials. The fact that SSP envisages a total 66,000 km length of canals including sub-minors (Desai and Joshi, 2008) poses an entirely new challenge to the irrigation engineers of the state. The biggest challenge comes from the fact that the canals carrying large volumes of water are likely to pass through some of the most water-scarce regions, but also agriculturally one of the most prosperous regions of the world—central and north Gujarat. This region has also been facing severe depletion of groundwater resources over the past several decades, and people here feel that getting more water from SSP is their legitimate right. Certain shocking realities with regard to regional agricultural trade enhance the legitimacy of these claims. Analysis done as far back as 2004, shows that north Gujarat, despite being absolutely water scarce, exported milk to other parts of the country including regions which are water-abundant. The water content of the milk was estimated to be 1,100 MCM per annum (Singh, 2004). North and central Gujarat, though water-scarce, have more arable land and can produce surplus agricultural output if irrigation water is provided. South Gujarat, though water-rich, has less arable land (Kumar and Singh, 2005).

A big challenge which SSNNL is facing today in making water available to the farmers of north Gujarat is in completing the construction of the network canals. Managing funds for constructing the canals is a major challenge, while the agency had developed innovative ways of making land acquisition procedure smooth. Lining of canals takes a major chunk of the construction cost. Hence, there are two conflicting interests. One is to provide greater volumes of water for meeting irrigation needs in central and north Gujarat, which are agriculturally prosperous, but facing water resource crisis. The other is to reduce the cost of building water distribution infrastructure.

What needs to be done to Maximise the Returns from Irrigation?

From the foregoing analysis it is clear that to maximise the economic, social and environmental return from the project calls for more judicious allocation of water from SSP among different regions to maximise the benefits of irrigation. Greater allocation of water to central and north Gujarat and Saurashtra, which are water-scarce, would produce incremental economic returns over what it would otherwise generate if that water is used up in water-rich south Gujarat districts, as illustrated by studies undertaken in western Punjab and eastern UP in the recent past which showed that the marginal return from irrigation is higher in water-scarce regions as compared to water-rich regions (Kumar *et al.*, 2008). Similar arguments were made by the renowned irrigation economist, B.D. Dhawan that marginal returns from irrigation would be high in low rainfall regions, and low in high rainfall region (Dhawan, 1990). Also, greater allocation of water, which can result in enhanced groundwater recharge in north and central Gujarat, would bring about high social returns, through improved sustainability of well irrigation, reduced energy cost for pumping groundwater owing to rise in groundwater levels, improvement in quality of groundwater and ecological benefits of improved flows in environmentally water-stressed rivers of north and central Gujarat. But this has to be done without any changes in the legitimate (volumetric) allocations of the farmers in different regions, as per the delta decided for each agro-climatic sub-region during the design phase.

Technical Approach

Water-scarce Areas of North and Central Gujarat

In the initial years, the irrigation dosage from SSP can be much higher than the delta determined (22 inches) by the Narmada Planning Group. This is possible, without changing the allocations earlier decided for each agro-climatic sub-region, in view of the fact that many of the large water projects planned in the MP part of the basin upstream of Sardar Sarovar dam, which would eventually help utilise all the dependable flows in the basin (Kumar, 2010) are yet to be taken up (Ranade and Kumar, 2004). The water, which is released from the upstream reservoirs during the rainy season and which is in excess of the storage capacity of Sardar Sarovar, which otherwise would flow into the ocean, can be diverted for irrigation through Narmada Main Canal. This can be done at almost zero marginal cost as the infrastructure for conveying the water will already be in place, and what needs to be done is to increase the duration of water supply from the canal system.

During these years, the maximum portion of water carried through the small canals in the rainy season (those below distributaries such as minors, sub-minors and field channels) and irrigation water applied in the field should be allowed to percolate down the un-saturated zone of north Gujarat's alluvial aquifers. This would help raise the groundwater table (Ranade and Kumar, 2004). Research studies undertaken in Murray basin of Australia and Indus basin in Pakistan, both having arid climate and deep water table conditions, showed substantial rise in water table conditions after the introduction of canal water for irrigation. Such changes happen due to reduced pumping of groundwater with the introduction of canal water, gradual increase in moisture storage in the unsaturated zone, which increase the unsaturated hydraulic conductivity of the soil media, and the increase in moisture pressure (hydraulic) gradient of the soil (Watt, 2008). While the first factor reduces or keeps the vertical distance for movement of recharge water, the second and third factors increase the rate of vertical movement of soil water (source: based on Richards' Equation and van Genuchten Equation as cited in Watt (2008: 101-102). This also means that from the point of view of economic efficiency lining the canal won't be a great strategy for the region which has deep groundwater table and which allows good percolation of irrigation water by virtue of the permeability. Nevertheless, it is understood that those canals which run on filling will have to be lined with bricks for preventing breaches and stabilising the embankments.

The strategy of releasing water during kharif season would help in maximising the positive externalities associated with introduction of canal water. Irrigation during the rainy season would increase the speed of vertical movement of water (return flow) through the unsaturated zone by virtue of the high hydraulic gradient of soil moisture, and higher hydraulic conductivity of soils in the unsaturated zone owing to higher degree of saturation, and low depth to groundwater. It will also increase the return flow portion of the water applied in the field due to high degree of saturation of the root zone soils. This would accelerate the recharge process. Since the opportunity cost of diverting the water will be insignificant, the benefits arising from improved groundwater recharge, groundwater quality remediation, improved quality of soils and rise in water levels could be treated as the additional economic benefits (positive externalities). The positive externalities include: expansion in irrigated area by wells and therefore the well irrigated production; reduced investment for well deepening by farmers; reduced cost of supplying water through alternative sources to villages and towns/cities affected by groundwater quality deterioration.

During the winter season, when the storage from the reservoir would be released for irrigation, the farmers can take up crops that give higher returns per unit volume of water, rather than those which are water-intensive and which give higher recharge per unit area of land irrigated. Nevertheless, it is important that a larger area is covered by canal irrigation during winter, as this would reduce the dependence on groundwater for irrigation in the region, thereby reducing the imbalance faster. For this the new command does not have to be developed through the construction of canal networks. Instead, water from the distributaries can be diverted to areas outside the designated command area for farmers to irrigate their land.

Once groundwater resources are built up in the over-exploited area, the canals can be lined depending on the groundwater level fluctuations in the area.[2] It is however, unlikely that the water level rise would cause waterlogging and salinity problems. The reason is that the total agricultural water demand in the region cannot be fully met from canal water, and as a result farmers would continue to depend on groundwater to meet their irrigation needs. The need for lining will not come so much from the fact that it is important for preventing water-logging and salinity, but from the fact that seepage would result in wastage of costly water. This is because the economic value of the benefits accrued from return flows would reduce gradually as groundwater situation in the region improves. But, such decision to

2. Some senior engineers of Sardar Sarovar Narmada Nigam Ltd. had expressed serious reservations about this. According to them, once the canals are left unlined for long periods of time, the channel regime would change. As a result, later on, there could be great practical difficulties in regaining the original channel shape for lining work to be taken up in future.

carry out lining should be based on empirical studies on the actual rate of seepage from canals. Research has shown that over the years, the infiltration losses from unlined canals stop permanently as the water table below the canal bed level raises due to continuous seepage and percolation saturating the entire underlying strata (Watt, 2008).

Water-rich Areas of South Gujarat

In the water-rich areas of south Gujarat, which are underlain by shallow and phreatic alluvial aquifers and shallow hard rock aquifers, the amount of water to be released for irrigation water supply will have to be limited. The reason is that excessive irrigation or irrigation of water-intensive crops such as paddy and sugarcane would lead to excessive return flows and finally, rise in groundwater levels. Since the groundwater balance in the region is still positive, this can create problems of water-logging and salinity in the command areas, similar to what happened in the Mahi command. If irrigation dosage from canals is limited, farmers could depend on groundwater in the area to supplement it. This would increase the opportunities for allocating greater quantities of water for north and central Gujarat, which receive lower quantum of rainfall, and experience higher aridity and experience groundwater depletion.

There is a strong economic rationale behind transferring a part of the entitlement of south Gujarat to north and central Gujarat. As studies show, the surplus value product from a unit volume of irrigation water is much higher in a water-scarce region as compared to a water-rich region. This is because the farmers in water-scarce regions allocate their water more efficiently to different crops, from a pure economic view point by putting more area under crops that give higher water productivity in economic terms and improving physical efficiency of water use. Also, the livelihood impacts of water used in irrigated agriculture are much higher than in unirrigated areas (Kumar *et al.*, 2008). Secondly, every unit of water transferred to north and central Gujarat for agriculture produces much larger social, ecological and environmental externalities, with great values realised in economic terms. Such positive externalities are either absent or are much less in South Gujarat.

The technical strategy for this would be lining the delivery canal systems, in addition to the distribution system. This would save the precious water, which otherwise would seep and percolate to shallow aquifers, creating waterlogging and salinity problems.

The Institutional Approach

The institutional approach to improving canal irrigation management in India had primarily focused on local institutional development, to enable farmer participation in management of irrigation systems (Brewer *et al.*, 1999). Market instruments such as volumetric pricing of canal water (Kumar and Singh, 2001), supply rationing (Kumar and Amarasinghe, 2009) and enforcement of volumetric water rights (Kumar and Singh, 2001), which are powerful tools for bringing about improvements in water use efficiency and effective demand management (Kumar and Singh, 2001; Kumar, 2010; Rosegrant and Schleyer, 1994 and 1996; Thobani, 1997) are not used (Kumar and van Dam, 2009; Kumar, 2010). The experiments in institutional alternatives had looked at various models, ranging from those which are aimed at managing irrigation affairs at the level of tertiary systems (Chambers, 1988; Merrey, 1996) to those which are meant for managing the entire hydraulic system starting from the main system to the tertiary system (Brewer *et al.*, 1999). However, the management functions envisaged for these local institutions included operation and maintenance of the system, water distribution, recovery of water charges, and conflict resolution. There is no dearth of examples from India of water users' associations enforcing water allocation norms amongst member farmers through measurement of volumetric water delivery from the inlet of the tertiary system, i.e., *chak*.

Pricing of Canal Water

Kumar and others had argued that volumetric pricing can be an efficient tool for water allocation in canal irrigation (Kumar and van Dam, 2009; Kumar, 2010) though many scholars have argued that there are theoretical and practical issues in using volumetric pricing as a tool for irrigation water demand regulation (Perry, 2001b; de Fraiture and Perry, 2002). In the absence of positive marginal cost of using canal water under the pricing structure that is based on crop area, the farmers in canal commands have no special incentive to use water efficiently. In fact, an earlier analysis carried out in the context of south Gujarat had shown that the implicit volumetric price for many water intensive crops such as sugarcane, banana and paddy, is much higher as compared to many low water consuming crops such as bajra and jowar (Kumar and Singh, 2001). Further, since there are no restrictions on the volume of water delivered owing to water abundance, the farmers in canal command areas of South Gujarat are found to be extensively growing certain water intensive crops that have high land use efficiency such as banana, sugarcane and paddy (IRMA/UNICEF, 2001).

Volumetric pricing of canal water would motivate farmers to use water efficiently in irrigating the crops, as it induces positive marginal cost of using water. They might still grow crops that are highly water intensive, but have water use efficiency (Kumar and Singh, 2001; Kumar, 2010). The reason is that they would be interested in maximising the returns from unit area of land, which is possible just by using more water under the present condition. This is because there are no effectively enforced restrictions on the volume of water that farmers can use (Kumar and Singh, 2001; Kumar and van Dam, 2009). It is important that farmers in central and north Gujarat use more water during the kharif season when the opportunity cost of using it would be low, but the recharge benefits relatively high. Therefore, it is important that the unit price of water is kept low for the water released during the kharif season.

Volumetric Water Entitlements in Canal Commands

After Frederick (1992), establishing privately-owned property rights that are tradable is critical to establishing conditions under which individuals will have opportunities and incentives to develop and use the resource efficiently, or transfer it to more efficient uses. If water entitlements of canal irrigators are fixed in volumetric terms, and if they are made tradable, regional water markets would emerge. The selling price of canal water is likely to go up in the market as demands are high. The demand for water is likely to come from naturally water-scarce regions of central and north Gujarat owing to two reasons: (a) the farmers there would be able to generate higher return from every unit of water used in farming; and (b) the area irrigated by SSP would be much smaller in comparison to the total arable land, and the potential of aquifers in the region to supply water to irrigate the remaining area is increasingly becoming poor. The water entitlement for the farmers in command area can be fixed in such a way that it creates artificial water shortage. In reality, this would anyway happen after the completion of the project under the Rotational Water Supply Scheme (RWSS) proposed for the project command, as the volume of water for each *chak* is decided on the basis of the irrigable area and crop water requirement for the main irrigation season, i.e. winter.

Well-defined tradable rights formalise and secure the existing water rights held by water users; induce water users to consider the full opportunity cost of water; and provide incentives for water users to internalise and reduce many of the negative externalities inherent in irrigation (Rosegrant and Binswanger, 1994; Kay *et al.*, 1997). If markets are efficient, the price at which water is traded would reflect the opportunity cost of using water (Howe *et al.*, 1986; Frederick, 1992). The price which north Gujarat farmers would be willing to pay for water would be the market price of water. Since groundwater markets already exist in the region, the price at which water is traded between well owners and water buyers can provide the price signals. Since the farmers in south Gujarat get low net returns from a unit volume of water used in irrigation for the dominant crops of the region such as banana, paddy and sugarcane, they would have a great incentive to shift to water-efficient crops, or use it efficiently or sell it at the market price and earn income (Frederick, 1992; Kumar and Singh, 2001). There is negligible scope for improving efficiency of use of water in the major crops of the region such as paddy and wheat under canal irrigation due to poor technical feasibility of using efficient irrigation technologies. Since the farming options that give high returns per unit volume of water (water productivity in economic terms) are limited in south Gujarat, farmers would be willing to transfer a share of their entitlement to central and north Gujarat and earn income. Given the high demand for water in north and central Gujarat, there would be large-scale water trading. The incentive for this would be high, as farmers do not have to invest in the infrastructure required for water transfer.

The Enabling Environment for Institution Building

If water pricing has to be introduced, measuring the volume of water used by farmers is a prerequisite. This has to be done at the point of inlet in the tertiary system and subsequently the water delivery at the field inlet of the individual farmers will have to be monitored. There are some notable examples of volumetric measurement of water delivery by WUAs at the inlet of the tertiary system. These are the Waghad and Mula irrigation projects in Maharashtra. On the other hand, if volumetric water entitlements have to be enforced, measurement of the volume of water used by the farmers has to be done for protecting the rights of individual water users and to make the markets work perfectly. This would also help monitor the physical transfer of water entitlements across regions, and check its legitimacy. For the first time in India, canal automation is being implemented in a large project like SSP. Since the hydraulic heads in the canal system would be continuously monitored and controlled, the volume of water delivered from each control point can be measured.

Since it would be impossible to work out water transfer agreements between hundreds of thousands of individual farmers located in two different regions, intermediate agencies will have to be created at various levels in the hydraulic system hierarchy—from main canal to branch and distributory canals. Transparency and accountability will have to be maintained by agencies that would be engaged in bulk water transfer at various levels in the hydraulic system. This would call for making the various levels in the hydraulic system independent, which means measurements of water delivery at various control points within the hydraulic system. This can also provide the basis of negotiation of water transfer agreements between parties, and payments for sale and purchase of water entitlements.

It is imperative that institutions will have to be created at various levels in the hydraulic system. The SSNNL had long ago proclaimed that irrigation water in SSP command will be supplied only to irrigation cooperatives or water users' associations and not to individual farmers. Hence, there already is an enabling policy environment in Gujarat. The need for such institutions would be quite felt by the farmers themselves, unlike in the case of WUAs being promoted in other canal command areas. This is because there would be strong economic incentives to get organised for the economic benefits that can be derived from access to irrigation water and specifically from water trading. In the past, the institution model promoted for management by farmers of irrigation in SSP had invited a lot of criticism from at least some quarters, arguing that it attracted huge transaction cost and that there was a sheer lack of interest among farmers in participating in irrigation management of the kind envisaged at the time of project planning (see Talati and Shah, 2004). But, if volumetric water pricing is practised and tradable water entitlements are enforced, farmers and other groups at various levels in the hydraulic system hierarchy would show great willingness to form organisations to engage in water trading at different scales as such transactions would yield returns.

This can unlock the real value of canal irrigation through SSP in Gujarat. Volumetric water transfer to the tune of 2,000 MCM to north Gujarat can earn farmers of south Gujarat and the intermediaries nearly Rs. 4 billion annually, if we assume that the farmers in north Gujarat pay Rs. 2 per m^3 of irrigation water purchased. This is quite reasonable in view of the fact that farmers from the region are paying more or less the same price for water from tube wells, which is of inferior quality due to the presence of salts. More importantly, as Table 1, based on a study carried out way back in 2003 shows, the net returns farmers in the region get from the use of water for crop production is far higher than this. But, the nature of institutions would be different. They should have capability to take care of the upkeep of the system, measure water volumes being transferred through various control points in the hydraulic system, and prevent illegal lifting of water from the distribution and delivery canals.

Table 1: Applied Water Productivity in Selected Crops in North Gujarat

Sl. No.	*Name of the Crop*	*Net Water Productivity of Crop (Rs./m^3) of Applied Water in north Gujarat*
1.	Kharif Castor	8.09
2.	Wheat	4.46
3.	Mustard	4.73
4.	Jowar	4.01
5.	Cumin	19.84
6.	Summer Bajra	2.85

Source: Based on Kumar (2005).

What is even more important is that in fact there would be significant economic gains for the intermediate institutions engaged in water trading owing to the fact that the physical transfer of water (which is to be traded between the regions) would not take place from the main system to the actual title owners of water in south Gujarat, once the trade agreements are negotiated between farmers in south Gujarat and those in central and north Gujarat. This would result in huge saving in water which would otherwise be lost in conveyance. The irrigation bureaucracy can give ownership of this water to the intermediate institutions, which would provide them strong monetary incentive to participate in water trading.

Finally, the monetary gains from water transactions should induce new incentives among the farmers below the minor outlets to form WUAs to carry out the larger irrigation management functions, which go beyond water transactions. Since such institutions will be self-driven rather than external agency-promoted, they are likely to sustain. That would also reduce the transaction cost of creating institutions.

Summary and Conclusions

The irrigation from SSP has the potential to produce multiple benefits. They include improved groundwater recharge and rise in groundwater levels in overexploited regions, enhanced inland fish production, enhanced biomass production in the villages, improved quality of groundwater for drinking and reduced carbon emission for power generation and improved health of riverine ecosystems. While some of them are unintended and now being perceived, some of them are even unforeseen. Given the fast changing water and energy environment in Gujarat, the economic value of these benefits can outweigh the direct benefits of irrigation, water supply and hydropower generation. The drivers of change in the economic dynamics of canal irrigation are groundwater depletion and its effects on energy as well as well irrigation economy, groundwater quality deterioration and the associated impacts on community health, environmental water stress experienced by rivers in central and north Gujarat and the destruction of ecological wealth. In order to maximise these benefits, irrigation water allocated for south Gujarat needs to be conserved, and more water needs to be allocated for central and north Gujarat. Enhanced water allocation for north and central Gujarat has to meet the needs of farmers, both within and outside the designated command area in that region. This can be done without changing the original water allocation plans and delta worked out for each agro climatic sub-region in the command, as un-utilised water from the upper catchment of the Narmada basin would be available for many years for diversion from Sardar Sarovar reservoir in addition to Gujarat's share of 9.75 MAF.

This would require institutional as well as technical interventions. South Gujarat needs to use SSP water more conservatively. Lining of delivery canals would be extremely important there, in order to reduce wastage of the water supplied and to prevent water-logging and salinity in the long run. Contrary to this in north Gujarat, enforcement of tradable water entitlements would motivate farmers in the region to sell part of their water entitlements to farmers of north and central Gujarat. Subsidised volumetric prices for water diverted through Narmada canal during kharif season would encourage farmers of north Gujarat to use it for crops that are water intensive crops. Trading of water can happen in bulk water transfer through the existing water conveyance infrastructure of SSP. Institutions are likely to come up at different levels across the hydraulic system hierarchy to engage in water trading, as they would benefit immensely in monetary terms through such trading. But, it is to be noted that unlike WUAs promoted at the tertiary system and above, these institutions would have high levels of management skills so that they can take care of the upkeep of the distribution/delivery system, measurement of volumetric water releases and preventing illegal lifting of water from the distribution and delivery systems.

29

Exploring Institutional Alternatives to Tap Fishing Potentials of Sardar Sarovar Reservoir

Samar K. Datta

Section 1: Background

1.1 Sardar Sarovar reservoir at Full Reservoir Level (FRL) covers almost 37000 ha of land area. This water formation has opened new avenues for fisheries development in and around the reservoir. As per the Narmada Water Dispute Tribunal (NWDT) Award, fishing rights of each state are limited to the water spread area within the administrative jurisdiction of each state. Fishing activities in this reservoir, therefore, need to be jointly carried out with equal participation by all the three party states.

1.2 Earlier the Government of Gujarat (GoG) had proposed formation of an Inter-State Fisheries Development Board subject to agreement by the party states to regulate development of fisheries in the reservoir. However, later in 1998, GoG mooted an idea to form a Joint Stock Company. The ideas of the other two states do not exactly match with Gujarat's. Moreover, whatever organisational format is chosen to conduct fisheries, it is necessary to have a clear idea about the proposed company's capital structure, the stake of each party state and cooperatives of project affected persons (PAPs), its organisational structure, Board of Directors, sharing of benefits, etc. Since choice of the final organisational format cannot be independent of these details, the High Level Expert Group of the Narmada Control Authority (NCA) on Fisheries Development decided to perform this exercise with the help of the present author to facilitate discussion and debate on this point. The present paper is an offshoot of this exercise.

1.3 Sardar Sarovar being a vast, indivisible and complex water resource with multiple dimensions and multiple stakeholders, the task of preparing a blueprint for fisheries development in this reservoir is conceptualised as one of brainstorming for evolution of a detailed system of property rights of the various stakeholders to this asset such that the various dimensions of this asset are properly conserved and developed to maximise its benefits. Japanese Corporations, which are believed to be the best-known examples of stakeholder cooperation (known as *keiretsu*) across multiple stakeholders (like shareholders, employees, managers, raw material and other order suppliers, wholesalers, retailers and even organised consumer groups) within the format of a joint stock company—a model which most of the corporate world is trying to emulate—thus provides an important conceptual framework for performing this exercise. Obviously, a large quantity of information was collected directly (through visits and interviews) or indirectly (from secondary sources—mostly from several relevant Departments under GoG) from various stakeholders and their spokesmen, besides processing it through conducting brainstorming sessions with them. At the beginning, a trip was undertaken to Narmada Control Authorities (NCA) headquarters at Indore for a preliminary interaction with the officials there. It was followed by three meeting-*cum*-brainstorming sessions—one each with the relevant officials in the Government of Maharashtra (GoM), Government of Madhya Pradesh (GoMP) and Government of India (GoI) in the presence of nodal officials from SSNNL (GoG) and NCA, besides a field trip to the Sarovar and one representative village to directly interact with the fringe population and oustees of the Sardar Sarovar Project (SSP).

1.4 The chapter is organised as follows: Section 2 spells out the technical details of the reservoir and its potential benefits, while Section 3 describes the concerns of relevant stakeholders. Section 4 performs a SWOT analysis of relevant organisational formats. Section 5 attempts to spell out the proposed organisational structure together with a roadmap for arriving at the 'idealistic' solution. The concluding section brings out the Coase Theorem conditions to explain why the reality may continue to elude and invalidate the value maximising proposition of this celebrated theorem.

Section 2: Technical Details of the Reservoir and its Fisheries Potential

2.1 Narmada is the fifth largest river in India and the largest west flowing river of the Indian peninsula originating from Maikala ranges at Amarkantak in Madhya Pradesh at an elevation of 900 m. It flows westwards over a length of 1,312 km before draining into the Gulf of Khambat, 50 km west of Bharuch city. The basin lies between east longitudes 72° 32′ and 81° 45′ and north latitudes 21° 20′ and 23° 45′. The Vindhya hills in the north, Satpura ranges in the south, Maikala ranges in the east and the Arabian sea in the west form the boundaries of the basin. The basin has an elongated shape almost like a thin ribbon with a maximum length of 953 km east to west and with a maximum width of 234 km north to south. The first 1,077 km of the river flows in Madhya Pradesh and the next 35 km stretch of the river forms the boundary between the states of Madhya Pradesh (MP) and Maharashtra. Again for the next 39 km, it forms the boundary between Maharashtra and Gujarat. The last stretch of 161 km lies in Gujarat. Out of the total catchment area of about 100,000 sq. km, 87 per cent lies in MP, 2 per cent in Maharashtra and 11 per cent in Gujarat. The technical details of the dam and the reservoir are listed below:

- Maximum height above the deepest foundation level (as at present): 163.00 m.
- Length of the reservoir: 214.00 km.
- Maximum width of the reservoir: 16.10 km.
- Average width of the reservoir: 1.77 km.
- Water spread area at full development: 34,867.00 ha.

The territorial distribution of the river is shown in Table 2.1.

However, the water spread area varies during the year, together with its distribution across the three states, as brought out by the following table from the GOPA consultancy report:

Table 2.1: Territorial Distribution of the Narmada River

State	*CultivableLand (Ha)*	*Forests (Ha)*	*Others (Ha)*	*Total (Ha)*
Madhya Pradesh	7,883	2,731	10,208	20,822
Maharashtra	1,519	6,489	1,592	9,599
Gujarat	1,817	4,166	1,069	7,112
Total	11,279	13,386	12,869	37,533

Source: Environment and Amelioration Programme (Fisheries), *SSNNL, Gandhinagar,* Jan. 2003, p. 6.

Table 2.2: Fluctuations in Water Spread Area under the Reservoir

States	*Minimum water spread in a year*	*Maximum water spread in a year*	*Average water spread in a year*
Gujarat	5,000 ha (45.5%)	7,469 ha (21.6%)	7,000 ha (30.4%)
Maharashtra	4,500 ha (40.9%)	6,570 ha (18.6%)	6,000 ha (26.1%)
Madhya Pradesh	1,500 ha (13.6%)	20,828 ha (59.8%)	10,000 ha (43.5%)
Total	11,000 ha	34,867 ha	23,000 ha

Source: GOPA Consultants—India: *Narmada River Basin Development Project (1991)*

Variation in depth along the reservoir together with the aforementioned fluctuation in water spread area across seasons within a year has clear implications for its fisheries potential as well as the claims of the three states on this potential—a matter which needs to be suitably addressed.

2.3 The judgment of the Chief Justice of India, the Honb'le B.N. Kirpal, dated October 18, 2000 in the writ petition (c) no. 319 of 1994, Narmada Bachao Andolan (Petitioner) *vs*. Union of India and Others (Respondents) in the Supreme Court of India highlighted the following expected benefits from the SSP:

(a) *Irrigation:* 17.92 lakh ha of land spread over 12 districts, 62 talukas and 3,393 villages (75% of which is drought prone areas) in Gujarat and 73,000 hectares in the arid areas of Barmer and Jalore districts in Rajasthan.

(b) *Drinking Water* facilities to 8,215 villages and 135 urban centres in Gujarat both within and outside command. These include 5,825 villages and 100 urban centres of Saurashtra and Kachchh which are outside the command. In addition, 881 villages affected due to high contents of fluoride will get potable water.

(c) *Power Generation:* 1450 Megawatts.

(d) *Annual Employment Potential:* 7 lakh man-years during construction and 6 lakh man-years after construction.

(e) *Flood Protection* to riverine reaches measuring 30,000 ha, 210 villages including Bharuch city and 7.5 lakh population.

(f) *Additional Benefits to:* (i) Dhumkhal Sloth Bear Sanctuary, (ii) Wild Ass Sanctuary in Little Rann of Kachchh, (iii) Black Buck National Park at Velavadar, (iv) Great Indian Bustard Sanctuary in Kachchh, and (v) Nal Sarovar Bird Sanctuary.

(g) *Development of Fisheries:* Deepening of all village tanks in the command which will increase their capacities, conserve water, will recharge ground water, save acquisition of costly lands for obtaining earth required for constructing canal banks, and reduce health hazards. Sophisticated communication system in the entire command area is also an important requirement.

2.4 However, considering the vast environmental implications of this project, the SSP was cleared from the environment angle on 24 June, 1987 by the Ministry of Environment and Forests (MoEF), GoI, subject to the following conditions:

- NCA will ensure that environment safeguard measures are planned and implemented *pari passu* with the progress of work on projects.
- Detailed surveys/studies assured will be carried out as per the schedule proposed and details made available to the Department for assessment.
- Catchment Area Treatment programme and Rehabilitation plans to be so drawn up as to be completed ahead of reservoir filling.
- The Department should be kept informed of the progress on various works periodically.
- Obviously, these are important norms which need to be fulfilled by any developmental activity involving the SSP.

Any proposal for development of fisheries in the SSP cannot be an exception to these rules.

2.5 Moreover, as many of the projects coming up on Narmada River and its tributaries involve submergence of forest land, Compensatory Afforestation works are required to be carried out on revenue and degraded forest lands in order to compensate for the loss of forests. However, fulfilment of targets in terms of compensatory afforestation doesn't automatically absolve any developmental plan on SSP of its responsibility to maintain and even promote the left over forest cover in and around the SSP. So, any plan to promote fisheries here must take this factor into account while developing the same.

2.6 It is important at this stage to highlight the role of Narmada Control Authority (NCA) which, following stipulations by the Narmada Water Disputes Tribunal (NWDT) and as per notification no. S.O.770(E) of the Ministry of Irrigation, was constituted by GoI on 10 September, 1980 to settle disputes between the states of MP, Maharashtra, Gujarat and Rajasthan on the utilisation of the waters of the Narmada River. It has been assigned the following six enumerated activities:

1. Storage, apportionment, regulation and control of the Narmada waters;
2. Sharing of power benefits from Sardar Sarovar Project;
3. Regulated releases by MP;
4. Acquisition by the concerned states for SSP of lands and properties to be submerged under the Sarovar;
5. Compensation and rehabilitation and settlement of oustees; and
6. Sharing of costs.

2.7 The NCA is thus a high powered body having the Secretary, Ministry of Water Resources (MoWR), GoI as its Chairperson, Secretaries in the Ministry of Power (MoP), MoEF, Ministry of Welfare (MoW), Chief Secretaries of the concerned four States (including Rajasthan, which is a beneficiary of irrigation from Narmada) as Members. In addition, there are a number of technical persons like Chief Engineers as the members. NCA is empowered to constitute one or more sub-committees and assign to them such of the functions and delegate such of its powers as it thinks fit. While NCA has enough responsibilities to discharge relating to its current activities, it seems it cannot do justice to a specialised technical function of fisheries development in SSP without directly involving the line departments in the respective states, given the socio-economic complexities of the local stakeholders, on the one hand, and interconnectedness of implied resources (especially water bodies) across alternative uses (agriculture, forestry, fishery etc.), on the other. So, before we proceed further, it is important to have a brief look at the stipulations of NWDT Award on fisheries development and the institutions directly or indirectly involved in fisheries development in the SSP.

2.8 Given both technical and socio-economic complexities in fisheries development over a vast water body, it is no surprise that the NWDT Award did not detail any formula for cost and benefit sharing in the context of fisheries, unlike what it could do for irrigation and power. However, it did quite pragmatically list down the following broad guidelines, which were to be operationalised by the involved states in active consultation with the multiple stakeholders:

- Notwithstanding the vesting in Gujarat of the lands coming under submergence, MP and Maharashtra shall continue to enjoy all rights of sovereignty over the submerged area in the respective States.
- MP and Maharashtra shall be exclusively entitled to all rights of fishing, boating and water transport over the part of the lake over the submerged land within MP and Maharashtra, respectively, provided however that such right is not exercised to prejudice of any utilities of the legitimate performance of their duties by the Project personnel.
- Studies related to downstream scenario for estimating impacts of the Project activities related to indenting of water for downstream by Gujarat.
- Rehabilitation of oustees from the submergence areas of Madhya Pradesh and Maharashtra, who are likely to settle in Gujarat or in their home states.[1]

1. *These stipulations are adapted from SSNNL* (2003, pp. 6-7).

2.9 In the absence of a clear-cut operational policy and/or coordinated efforts towards fisheries development in the SSP, the three relevant states seem to have followed altogether different courses of action so far. While the state of MP, holding the lion's share of this reservoir, has been pursuing fisheries independently within its physical jurisdiction with the help of local fishermen organised under the State Fisheries Federation and under a contractor, Gujarat and Maharashtra are yet to draw up an operational plan, though half-hearted and informal fishing seems to be taking place in isolated local patches within the geographical boundaries of these two states. Both the states of Gujarat and MP claim to have been regularly stocking fingerlings in the reservoir, though separate figures for stocking in the reservoir are available only from the state of Gujarat. The Government of Maharashtra too has come up with an action plan and budgetary provision to stock and promote fisheries in the reservoir.

2.10 Table 2.3 lists the institutions, that are currently involved or can be potentially involved in promotion of fisheries in the SSP. It can be seen from this table that, except for NCA, no other organisation has the mandate to resolve possible inter-state differences in approach towards fisheries development, thus pointing to a possible institutional vacuum to achieve the purpose in hand. Although the interests and stakes of these institutions vary quite a lot, they can nevertheless be looked upon as institutional stakeholders and thus can be suitably networked for promotion of fisheries in this water body.

2.11 Before we explore institutional alternatives to tap the potential of fisheries in this reservoir, it is important to get a reasonable estimate of the size of the carrot, which can create a suitable incentive structure for the stakeholders to join hands with one another and make it profitable for them to incur the necessary cost of developing and promoting an organisational format to reap the benefits. Unfortunately no precise fish productivity figures are available for this reservoir in the absence of any systematic effort to pursue fisheries in it. Coming up with a good estimate has also been constrained by the fact that no detailed data—whether cross-sectional or inter-temporal—are available on the technical features as well as fisheries management practices on reservoir fisheries, so that one can project reasonable estimates of yield for any particular reservoir on the basis of its observed technical features and fisheries management practices

Table 2.3: Institutions Currently Involved in Development of Fisheries in SSP

Institutions	*BroadFunctions*
Narmada Control Authority (NCA)	Nodal inter-state control authority responsible for coordinating activities around the Narmada river. Though not a fisheries institution, it can be of immense help in coordinating fishing activities of the three states.
Narmada Valley Development Authority (NVDA)	Inter-departmental agency of the Government of MP, coordinating all activities related to the Narmada basin.
Sardar Sarovar Narmada Nigam Limited (SSNNL)	Counterpart of NVDA for the state of Gujarat.
Fish Farmers' Development Agency (FFDA)	Created by Government of India in 1973 to promote intensive fish culture all over India. Though all districts of MP, Maharashtra and Gujarat seem to have been covered by FFDA, Sardar Sarovar is yet to enjoy its benefits.
Marketing institutions	Predominantly private middlemen, though informal fishing in the reservoir may be performed by some predominantly state-owned cooperative bodies. The weakest link in the value chain.
Research	Central Inland Capture Fisheries Institute (CICFRI), Barrackpore, West Bengal, an ICAR organisation, with branches in several states, is mainly engaged in providing research inputs.

(including stocking rate). In this situation, all we can do is to come up with a conjectural figure on the basis of the following feasibility figures of GOPA Consultants (1991) and the observed yield figures (as reported in Tables 2.4 and 2.5 respectively).

2.12 A rough work on the net value of fish production from the SSP is done in Table 2.6 under three alternative sets of assumptions about yield rate in large reservoirs—(i) prevailing average yield rate of 11 kg/ha/year with hardly any stocking; (ii) an average potential yield rate of 50 kg/ha/year, as assumed by the National Commission on Farmers (2005) assuming a normative stocking rate of 300/ha; and (iii) a maximum potential yield rate of 200 kg/ha/year corresponding to a stocking rate of 300/ha, as realised in some of the reservoirs—including the Ukai reservoir—in Gujarat under a fairly rigorous fisheries management system. Thus, assuming average yearly water spread area of 23,000 ha under the SSP, we get a net annual income figure of about Rs. 5 million in the most conservative scenario, about Rs. 12.7 million in a more liberal scenario, and about Rs. 81.7 million in the most liberal scenario.

Table 2.4: Fisheries Production and Income Potential of SSP[2]

- 40 kg of marketable fish/ha of average water spread in the first three years following impoundment (ending June 30, 1997) and 50 kg in the subsequent two years (ending June 30, 1999).
- Figures based on empirical data and information available in literature, besides morphoedaphic features of the reservoir. It is assumed that a comprehensive culture-based fisheries management plan with regular stocking, initial stocking adjusted to trophic burst, ban on fishing for first two years, fishing within the sustainable yield etc. will be implemented.
- Assuming a catch of one ton of marketable fish per year per fisherman, whereas fisherman's per fishing day consumption is 1 kg. (mostly smaller sized and non-marketable fish) and price of fish is Rs. 15/- per kg., a fisherman can earn an annual fisheries income of Rs. 15,000. The GOPA Report felt this income level was enough to meet the fisherman's expenses for nets and gear besides enjoying a standard of living reasonably above the 'poverty limit' set by Indian authorities.
- The GOPA Report's financial analysis reported 17 per cent economic rate of return, even after leaving out additional income generating activities.
- The GOPA Report's financial analysis reported 17 per cent economic rate of return, even after leaving out additional income generating activities.

Source: *GOPA Consultants* (1991), pp. 3, 4 and 15 of Annex 5.

2.13 Obviously, the lower limit on income generation capacity of this rather untested[3] reservoir puts a drag on the institutional alternatives one can pursue in this context. This is especially true of the Joint Stock Company alternative where, given the current low yield rate and

2. GOPA Consultants (1991) also provide some data on fish catch from the reservoir (pp. 3-4, Annex 3) "….Interpretation of scanty 1960 data on (most probably marketed) fish production and number of fishing units on the main river channel in MP leads to a low yearly catch of 30 kg/fisherman or 40 kg/fishing boat. Interpretation of 1988 data from various sources on the specialised professional fishery in the Narmada estuary however show a very reasonable catch per unit effort of 2600 kg per fisherman/year or 9500 kg per fishing boat/year." Moreover, Parikh (1986) as quoted by GOPA shows annual net income per fishing unit from Bharuch district in Gujarat as Rs. 5,964 from the river and Rs. 10,515 from the estuary.
3. Rather untested because fishing is yet to be systematically pursued on this reservoir.

Table 2.5: Current vis-à-vis Potential Yield Rates from Indian Reservoirs

(*Figures in kg/ha/annum*)

Category of reservoir	*Average yield*	*Potential yield*
Small (<1000 ha)	49.90	100.00
Medium (1000-5000 ha)	12.30	75.00
Large (>5000 ha)	11.43	50.00
All	29.70	77.00
Govind Sagar	120.00	–

Source: National Commission on Farmers, Ministry of Agriculture, GoI, August, 2005.

Ukai (60060 ha) (Dist: Surat)	208.28	–
Panam (Dist: Panchmahal)	245.91	–
Dharoi (Dist: Mehsana)	214.91	–

Source: Government of Gujarat.

Alternative Scenarios	Catch value @ Rs. 25/kg	Cost of catch @ Rs. 5/kg	Stocking cost @ Rs. 1.5/piece[4]	Income net (Rs)
I. Most conservative	275	55	negligible	220
II. Average potential	1250	250	450	550
III. Maximum potential	5000	1000	450	3550

Source: National Commission on Farmers, Ministry of Agriculture, GoI, August, 2005.

uncertainties over the rate of return on this asset, no investor—whether private or public—is likely to feel attracted to invest in this proposition, at least for the time being, though it is quite natural for GoG to dream of the joint stock company alternative, given its successful achievements in some of the reservoir fisheries within the state. So, in this situation, given an uncertain and in all probability a low return from this alternative, it is understandable why some of the state governments are keen on pursuing the cooperative alternative and vesting of fishing rights in the local fishermen. However, there are obvious dangers in being too conservative or too optimistic on fisheries development in the SSP at this stage. If only the alternative of cooperative of exclusive fishermen is stuck to, given the poor record of cooperatives in this

4. Market price of Rs.25/kg, cost of catch of Rs.5/kg and cost of quality fingerling of Rs.1.50/piece are assumed following discussions with experienced GoG and GoMP officials.

country in general and in the context of fisheries in particular, this may be the end of the tunnel, ruling out significant developments in technology and institutional innovations to generate enough value out of this asset to improve the fish farmers' lot[5]. The joint stock company option, on the other hand, may not be lucrative to start with, but given the technological and especially institutional innovations around the world in reservoir fisheries, there is no reason why one should stop dreaming of this alternative for all time to come. Given the emerging trend towards stakeholder cooperation in the management of complex assets with multiple stakeholders, it is possible to combine the advantages of a farmers' cooperative solution in the backward linkages of the value chain with the advantages of a corporate approach in the forward linkages component, namely in post-harvest management, and to gradually and in a planned manner drift towards a corporate approach for realising the highest possible value out of this resource. Before we spell out the details of this approach together with a road map in the sections which follow, it is important to emphasise that one should not be bogged down even by the liberal figures quoted in the preceding table. So, we would like to end this section with a note of optimism, highlighting various technological alternatives to generate more value and income through an integrated approach to fisheries development in the SSP.

2.14 The maximum potential income from fishery activities from the SSP need not be the maximum for all time to come, not even in the near future, if innovative technological practices are followed to achieve value addition and diversification of activities are undertaken to tap economies of scope on the same resource base. We briefly mention below only some of these options:

- Yield can be augmented substantially by undertaking cage and pen culture, as is being practised in the Tungabhadra reservoir with modest amount of success and as recommended by GOPA consultants in 1991;

---

5. In MP, yield rates are found to be considerably higher in Department-managed water bodies than in Apex Cooperative-managed bodies. Moreover, yield rate is found to be strongly positively correlated with the stocking rate.

- Yield in value terms can be significantly improved by changing the composition of fish culture in favour of high valued species. As SSNNL (2003) has pointed out, it is even possible to undertake production of Hilsa[6], giant freshwater prawn (Table 2.6 provides information on existing scampi culture in areas near the dam in Gujarat) and Mahseer at least on a limited scale using appropriate conservation measures towards these vulnerable species;
- More income and employment opportunities for the local population, which is a must if the fates of these people are to be improved alongside sustainability and even progressive improvement of the SSP reservoir fisheries, can be created through undertaking seed production *in situ*, net-making, boat-making and repair-*cum*-maintenance activities, transportation and storage activities;
- The fringe population can be suitably organised through formation of self-help groups under existing schemes of the Ministries of Rural Development, Tribal Welfare, Environment and Forests etc. and utilised in conservation-*cum*-ecodevelopment activities to tap economies of scope and generate supplementary sources of income for them, so that a healthy interface can be created between the fringe population and the resource called SSP for mutual development; and
- Instead of pushing the fringe population towards value-adding forward linkage activities to begin with, a pragmatic step would be to follow the example of the National Dairy

6. SSNNL(2003) on p. 15 has made the following observations on Hilsa culture: "The first ever attempt to breed Hilsa of the Narmada Estuary was carried out by Malhotra, *et al.* (1979), who had previously deployed hypothysation and other breeding techniques on the Hilsa elsewhere (1969 and 1973). Malhotra *et al.* (1979) could transplant this species into the nearby Vallabh Sagar Reservoir at Ukai (Surat district), in southern Gujarat. It is worth mentioning in this connection, that Hilsa has established itself within the Ukai Reservoir and is reflected in the fish catch of the reservoir. The size of the fish, however, is reported to be small, when compared to the marine forms, and the new strain seems to have adapted to fresh water environment."

Development Board (NDDB)[7] to involve good corporate business with strong commitment and proven records in rural development for harvesting, processing and marketing of fish on contract basis, but in a spirit of private-public partnership at least in the initial stage of development of fisheries in this reservoir.[8]

Table 2.6: Information on Scampi Culture in Gujarat

District	*Extent of Farming (Ha)*	*Estimated Catch (Tonne)*
Valsad	722	29
Surat	630	295
Bharuch	590	350
Vadodara	85	185
Kheda	444	200
Ahmedabad	106	13
Surendranagar	3061	200

Source. *Environment and Amelioration Programme (Fisheries),* SSNNL, Gandhinagar, Jan. 2003, p. 30.

Section 3: Concerns of Relevant Stakeholders

3.1 The purpose of this section is to identify the major stakeholders in the SSP and to bring out their concerns and aspirations, so that an appropriate organisational format can be evolved through negotiations across these stakeholders. Three broad categories of stakeholder are identified—(i) the concerned Ministries/Departments of GoI, which have cleared this project under certain stipulations, and the scientific

7. It is learnt in course of discussion with NDDB officials that NDDB had taken help of a private corporate enterprise called Voltas in marketing of milk products of the dairy cooperatives at the initial stage before the brand name of AMUL became a household name throughout the country and even abroad.
8. It is precisely because of the above-stated potential, GOPA Consultants had proposed for possible World Bank funding a package of investment measures inclusive of (i) Construction of 8-ha hatchery complex, (ii) Provision of rearing facilities (rearing ponds, pen and cage culture), (iii) Provision of infrastructure for landing, storing, handling, and marketing the catch, (iv) Enforcement of fishing regulations, and (v) Technical Assistance.

communities, which came to be associated with them, (ii) the project affected families and the fringe population—whether actual or potential fishermen on full-time/part-time basis; and (iii) the three state governments—GoMP, GoM and GoG, the joint claimants to fisheries in this reservoir.

3.2 MoEF had stipulated implementation of the following[9] environmental safeguard measures *pari passu* with progress of the engineering works of the various projects including SSP on the Narmada river:

- Phased Catchment Area Treatment
- Compensatory Afforestation
- Command Area Development
- Survey of Flora and Fauna and Carrying capacity of surrounding areas
- Health Aspects
- Pisciculture
- Archaeological and Anthropological studies
- Seismicity and Rim Stability of Reservoirs.

The additional parameters on which various other Departments of GoI have expressed concern relate to Rehabilitation Master Plan and Downstream Impact (SSNNL, 2003: 7-8).

3.3 While NCA has been monitoring progress on almost all these parameters in consultation with other relevant bodies of the central government as well as the concerned state governments, the progress on tapping the potential of fisheries in SSP has been extremely slow, apparently for several reasons. First, fisheries is a complicated subject with uncertain outcome, especially when a huge common property resource is involved with poorly defined property rights of a multiplicity of stakeholders including a poor and backward fisher folk community that was too often divided by a difficult terrain on both sides of the reservoir. Second, the negative externalities of practising fisheries in this reservoir, often highlighted in the literature, seem to have downplayed the significance of fisheries in general, and undermined the enthusiasm of its promoters, while the negative impact (e.g. pollution) of other activities, which the champions of fisheries have usually failed to control, has further dampened their interest.

9. *Source:* http://www.nvda.nic.in/env.htm

Third, clearance from MoEF, which is a pre-condition for any organised fisheries activity in SSP, seems to have acted as another barrier—especially in the absence of any well-established tradition or a common platform for coordination of activities between the two relevant Departments/Ministries. The term 'environment' becoming associated with forestry, rather than remaining a neutral common platform of several concerned Ministries/ Departments has often become an overriding concern to the utter neglect of the potentials of fisheries. Finally, the three state Fisheries Departments too, have failed to see eye to eye with one another to develop a common platform and plan, besides a common set of strategies. Nor could they optimally pursue independent fishing on this contiguous water body spread over a very large land area cutting across several states. It is precisely because of these complications and possible hurdles; the GOPA Consultants (1991) in their report to the World Bank on the potentials of fisheries in SSP had highlighted the need for fresh studies and rigorous monitoring on a number of issues, as detailed in Tables 3.1 and 3.2, respectively. Obviously, the question is, who will bell the cat, in the absence of a suitable incentive/ disincentive structure? Needless to add, in the absence of the necessary property rights (i.e., incentives and accountability), these tasks tend to be undertaken inadequately, if at all, as has been the case in practice.

Table 3.1: Studies Required on Sardar Sarovar Reservoir as per GOPA Report

1. Study on natural habitats of commercial fish
2. Reproduction patterns and growth of commercial fish
3. Feeding habits of commercial fish
4. Bio-productivity of aquatic ecosystems
5. Experiments with cage and pen culture techniques
6. Experimental fishing

3.4 Exactly the same question arises with respect to management of health, hygiene and sanitation of the fringe (and mostly resettled) population, a large chunk of which practises fisheries in the SSP—whether formally or informally and whether full-time or part-time. Although Line Departments of the relevant states are discharging these health-care functions, these are being performed in the spirit of obligations and duties with outside funding, rather than as claims by the beneficiaries. If most of the project affected people (PAPs) are practising fishermen, it will be logical to put the concerned Fisheries

Table 3.2: Monitoring Requirements on Fisheries in the SSP as per GOPA Report

1. Physico-chemical parameters
2. Biotic parameters
3. Ichthyological monitoring
4. Fisheries statistics
5. Development of waterlogged areas
6. Bio-monitoring of waste-water discharge points in the estuary
7. Mangrove reforestation
8. Development of waterlogged areas

Departments at the centre-stage of the responsibility system for discharge of these functions. If such people are contributing to the revenues of the state Fisheries Departments, not only shall the Departments have the urge to take care of the health needs of these people, but the people can also suitably pay for such services in the foreseeable future, even if the Departments procure these services from other Departments for onward supply to these people. On these health problems, GOPA Consultants (1991) have made the following remark, which is worth noting and clearly makes a case for integrated fisheries management—that is, taking care of the urgent concerns of the fishermen while promoting fisheries:

> "In general, malnutrition does not seem to be a problem. ....diseases are more related to insufficient water quality and availability in hilly areas outside the submergence zone and stagnant and/or polluted water in the plains and estuary with a high rate of skin problems reportedly due to poor hygiene and intestinal diseases due to bad water" (p. 9, Annex 3).

3.5 Three different bodies—namely MS University of Baroda (1983), CICFRI, Barrackpore (1993) and Wallingford (1995), have expressed ecological concerns over fisheries development in the SSP, besides suggesting mitigating measures. A close look into these suggestions reveals the following:

- Some of the remedial measures suggested—for example, pumping out unsuitable ground water, selection of crops—go beyond the scope of activities of a Fisheries Department.

However, these issues have been raised whenever any serious thoughts are given to tapping the long-untapped fisheries potential of this reservoir. Naturally, an exclusive body of Fisheries Departments cannot undertake these tasks—both logically and due to financial constraints. Hence, if the above-stated scientific advice is to be adhered to, the proposed institutional body must have a broader mandate and a larger resource as well as support base (with close coordination with several non-fisheries departments) to pursue fisheries in the SSP in an integrated manner.

- A huge amount of R&D inputs will be needed almost continuously to undertake the necessary remedial measures, which call for massive investment including investment in networking with suitable R&D bodies across the country and even beyond, from the promoters of fisheries in the SSP.
- Given the vastness of the resource, no governmental body would be both technically and financially equipped enough to monitor the prescribed ecological parameters in a continuous fashion, unless a participatory approach involving support and cooperation of the fringe population—irrespective of whether they are fishermen or not—is assured. Several State Forest Departments, for example, have been attempting to ensure forest protection and sustainable forest development through local forest protection/eco-development committees.
- Concern over vulnerable species including Hilsa, as expressed by certain quarters, may be overstated. As an SSNNL Report (2003) has brought out (pp. 23-34), with suitable conservation measures underway it is possible to drag down the extent of loss to a minimum level. However, the proposed body must be both technically and financially strong enough to implement the proposed conservation measures.
- Last but not the least, given the fact that there are about 1300 perennial village ponds and another 40,000 ha of waterlogged area around this reservoir with strong fishing potential, it is quite natural for the fringe population to dream of fishing in these water bodies. The question is whether the proposed

organisational set up would concentrate merely on fishing in the reservoir as per its mandate or would lend support to the fringe fisherfolk to develop these water bodies as well. The GOPA Consultants (1991) are in favour of keeping this matter altogether separate from development of fisheries in the SSP, although it is always possible to devise an organisational format which, while concentrating on the reservoir, can also address the issues of village ponds and other water bodies as a gesture of goodwill to the fringe population, particularly when promotion of fisheries on this vast reservoir in any manner whatsoever is an impossible proposition without active involvement and support of the fringe population.

3.6 We now turn to the socio-economic concerns of PAFs and fringe population. Fishing is an activity where the act of production/catch itself involves direct and intense contact with the fisher folk, unlike in other activities like providing canal irrigation or supplying power. So, in the absence of any concerted effort to promote fishing in the SSP, the fishermen and fisherwomen around this reservoir as well as their multifarious socio-economic problems seem to have gone into oblivion. It is, therefore, no surprise that there is very scanty amount of material available on the socio-economic status of the PAFs and fringe population. Even though some information is available, it refers to the pre-inundation period. The conspicuous absence of any panel data on the socio-economic conditions of the PAFs at different points in time makes it impossible to trace their growth and development, not to talk of designing plans and programmes to suit their aspirations and dreams. It appears the three state governments maintain records of these families (maybe even in computerised form), but there is hardly any record of or even attempt to record their dynamics over time[10]. So, in this situation, we have proceeded on the basis of whatever little bit of data is available on the PAFs from

10. This is sharp contrast to the records on wildlife, which, we fear alongside a very senior official of GoMP, may facilitate establishment of wild life sanctuaries in certain stretches of the reservoir at the cost of the fisher folk. Absence of documentation thus hurts the cause of the fishermen not merely because of lack of possible hard evidence,

existing websites, from a socio-economic survey by CICFRI (1993), from GOPA Consultants' Report (1991) and from a field visit to a representative village in the fringe area within the state of Gujarat. We list below some of the findings from these reports, which seem to have important bearing on the task in hand:

- "The most vulnerable among the affected are the fisher folk, who depend on riverine fisheries for a living. Unorganised and tradition-bound, the artisanal riverine fishermen in the country live either at a subsistence level or below the poverty line. Construction of dams radically transforms the lotic environment and robs the fishermen of their daily bread. Since these fishermen know no other vocation, their rehabilitation needs careful planning and consideration" (*ibid*, p. III).
- As many as 66.25 per cent of these families are landless, whereas 4.92 per cent have less than 1 ha of land, 4.93 per cent have 1-2 ha of land and 3.97 per cent more than 2 ha of land.[11]
- The average number of active fishermen and fisherwomen are found to be 2.9 and 2.56, respectively, per sample family.
- The percentages of scheduled caste, scheduled tribe and other families are found to be 1.5, 31.0 and 63.6, respectively.[12]
- Of the total families, 62.1 per cent belong to the fishermen community, whereas 29.9 per cent belong to other communities.[13]
- The average number of livestock per household is 1.17 heads
- While the womenfolk of the sample families spend most of their time (70.64%) in domestic affairs, they spend 11.65 per cent, 0.87 per cent, 0.24 per cent respectively of their time in agriculture, livestock maintenance and outside work The extent of their time spent on fishing, fish processing

---

but also because such a situation facilitates plight of the mostly unorganised fishermen in this inland segment in an unsung and unheard manner without providing any natural justice to them.

11. Of those surveyed, 13.44 per cent do not report details of landholdings.
12. The remaining 3.9 per cent haven't responded.
13. The extent of non-response is 8 per cent.

and fish transport is 10.80 per cent, whereas 0.19 per cent, 0.56 per cent and 0.03 per cent of their time are spent respectively on net making, retail fish marketing and wholesale fish marketing.[14] A large percentage of these people are reported as keen to receive vocational training.

- The average fish catch of these families is reported as 875.4 kg per family per annum, the bulk of which (45% and 37.2%) is sold respectively in retail and wholesale markets.
- About 4.8 per cent of the fishing families release the young fish back into the river for further growth, and only 14.3 per cent families observe a closed fishing season during the fish breeding period.
- Only 15.1 per cent of these families migrate locally in search of better fishing grounds.
- Almost all of them use traditional fishing craft and gear.
- Among the fishermen, 54.3 per cent are earning members and 45.1 per cent non-earning members. The corresponding figures for fisherwomen are 28.2 per cent and 70.6 per cent, respectively.
- The major occupational diversification observed among the sample fishermen and fisherwomen is reported in Table 3.3 below. The incidence of subsidiary occupations is found to be 37.04 per cent among fishermen and 19.81 per cent among fisherwomen.
- As many as 91.3 per cent families have expressed their interest in fishing in the reservoir, when it would be created and 91.7 per cent of them have expressed interest in aquaculture training. (*ibid,* pp. IV-X).

3.7 As the above-stated reports provide a socio-economic perspective of the PAFs and fringe population only before inundation, we decided to undertake a fresh field study in the form of a PRA exercise in a village named Suka near the dam site in Gujarat during September, 2005. The findings are narrated in the form of a short story in Appendix 1 to provide at least a qualitative perspective of

14. The remaining 4.97 per cent of time is spent on miscellaneous activities.

Table 3.3: Major Occupational Composition of Fishermen and Fisherwomen

Occupation	*Major Occupation*	
	Fishermen (%)	*Fisherwomen (%)*
Fisheries	88.56	32.08
Agriculture	2.95	24.11
Livestock	0.04	0.00
Labour	5.28	33.47
Service	1.34	0.13
Teaching	0.07	0.00
Cloth washing	0.01	0.00
Woodcutting	0.04	0.15
Dairy	0.32	1.32
Others	0.74	4.34
No response	0.37	2.26

Source: *CICFRI* (1993), p. VIII.

the socio-economic problems of the PAFs and fringe population, which ought to be taken into consideration while designing a suitable organisational format to tap the fisheries potential of the SSP. A few important lessons to note in this context are:

- Although there has been a considerable diversification in occupational pattern, there is still a fair amount of interest in and dependence on fisheries in the reservoir. However, the PAFs need updated equipment and training, besides support in forward linkage activities (especially in storage, transport and marketing) to cope with the new situation.
- Given low and rather uncertain outcome of fisheries in the reservoir, at least in the initial stage, these people need to be encouraged and empowered through training to undertake supplementary and even alternative sources of income in a planned manner, possibly under an SHG approach, rather than under an outright cooperative structure imposed, capitalised and controlled from outside. The active work involvement of women in both pre-and post-inundation stages raises considerable hope for SHGs, which can tap supplementary resources from various existing governmental schemes of different Departments/Ministries and, at the same time, begin to meet at least part of their credit needs internally.
- Although several government Departments have been

meeting the PAFs' infrastructure and social overhead requirements in the rehabilitated villages till now, this is unlikely to be a sustainable solution. Hence, once organised fisheries activities start under a suitable organisational format, the latter would not be in a position to escape the responsibility of sustaining the villagers' infrastructure requirements. Hence there is a vital need for an integrated approach concentrating on the fisher folk and their multifarious requirements, rather than merely on fisheries as a single activity.[15]

- Since most PAFs and fringe households pursue several activities including fisheries simultaneously at different parts of the year as a risk-hedging device, it is very difficult to make a watertight distinction between fisher folk and non-fisher folk households. Moreover, managing and monitoring a vast resource like the SSP is not simply possible without the support of local people, even if they are engaged in allied non-fishing activities—be it agriculture, forestry, livestock, whatever. Hence, there is need for coordination and integration across several allied activities, generally looked after by several different government Departments. Obviously, the proposed organisational format must acquire the flexibility and capacity to follow this integrated approach.

3.8 Finally, we turn to the concerns and proposals of the concerned states. Both GoG and GoM are agreeable to formation of an interstate body with equal representation of the three states and nomination of its Chairman by rotation. However, GoMP did not agree to this proposal and proposed a Board under its control. It argued as follows:

---

15. It is probably worthwhile at this juncture to take note of the infrastructural position of the fringe villages, as narrated by the GOPA Consultants (1991): "Villages of tribal fisher folk do not have sufficient roads in sufficient, transport, electricity, proper health care and market facilities. Low rate of literacy is mainly due to the non-availability of school facilities in many tribal areas before 1987. It is evident that these villages have been neglected all along for a variety of reasons, but mostly due to geographical isolation."

> " ...in view of the fact that nearly 60 per cent of the total area under submergence will be within Madhya Pradesh territory, the management of SSP reservoir for fisheries should be given to this state. This is also in consonance with the proclaimed state policy which states that the inter-state reservoirs (i.e., reservoirs which are on common borders) should be developed by the state in whose jurisdiction the greater part of the reservoir falls (i.e., single control).
> It is also proposed that for the integrated development of the reservoir, monitoring and evaluation etc. a committee of direction may be constituted under the NVDA, which will also be represented by a nominee from each of the states of Gujarat and Maharashtra. The said representatives should also ensure that the dictates of NVDA are fulfilled.
> Further, the Gujarat Government should agree to bear the financial cost on development of fisheries, as this activity constitutes an integral part of the overall environment development programme, to which the Gujarat Government is committed under the environmental and other clearances by Government of India to Sardar Sarovar Project." (Vice Chairman, NVDA's letter No. NVDA/D (Fish)/88/3867 of 1/11/88 addressed to the Chairman, NCA)

3.9 However, as recorded in a document titled 'High Level Expert Group on Fisheries Development and Conservation (p. 7),

> "Gujarat ... disagreed to the proposal of formation of Apex Cooperative Federation, where states are represented through district level user cooperatives as only a part of the district Dhule in Maharashtra and part of a district Narmada in Gujarat but three districts—namely Barwani, Jhabua and Dhar of Madhya Pradesh shall be involved. Gujarat was also not willing to grant complete rights of fishing in the entire reservoir to MP."

3.10 Consistent with its earlier stand alongside GoG, GoM developed a detailed proposal for establishment of a Narmada Inter-State Fisheries Development Authority (NIFDA) and forwarded it to the Secretary, NCA on 28/10/1991, but apparently in view of the unchanged stand of GoMP for a Board under its control, GoM in a D.O. letter No. Matsyavi-2491/CR-236/ADF-13 of 28th March, 1994 from Agriculture, Animal Husbandry, Dairy Development and Fisheries Department to Deputy Secretary, Narmada Ghati Development Department, GoMP conveyed its approval to nominate Commissioner for Fisheries, Maharashtra State, Bombay as member on Sardar Sarovar Reservoir Inter-state Fisheries Board. However, at a later stage—precisely on 2/9/05, Commissioner of Fisheries, GoM withdrew this nomination arguing that they "do not intend to become a Member of your Board constituted by GoMP" (Reference: Their letter No. Fish/T/011403/2168 to Secretary, Narmada Ghati Development Department, GoMP).

3.11 Meanwhile, probably considering the stalemate over the three party states' agreement to a common organisational format to tap the potential of fisheries in the SSP, GoMP floated an 'Alternative Proposal', which was forwarded to GoG by NCA on 26/11/1996. In the last paragraph of the said document, GoMP argued as follows:

> "..........In the past, the GoMP had suggested an Apex Cooperative Federation for fisheries development and management in the Sardar Sarovar reservoir. This could be modified to include private sector participation also in the light of subsequent developments regarding project financing. Private enterprise in this sector is, however, still an untried proposition, but may be given a trial in view of the emerging interest in this area. Alternatively, therefore, a joint-sector venture can be conceived with proportionate shares (in the majority stock-holding) by the party States in the initial instance, to be diluted in favour of the private promoter after the venture has achieved viability. The advantage in such an arrangement would be that while the State's investment/developmental aims could progress on economic considerations, they could be left free to address themselves to the important concerns of training fishermen and rehabilitation/resettlement of displaced persons. The social commitment regarding fishermen's welfare could be codified into the Articles of Association/Company Policy. A

joint-stock company with government share-holding may also allay misgivings regarding grant of long-term lease to non-government agencies and apprehensions regarding extinction of bio-diversity; and at the same time enable a dominant voice for the larger share-holders if investment is forthcoming."

3.12 Apparently considering the vast potential of a diversified set of activities around fisheries in the SSP, on the one hand, and its experiences of enhanced fish productivity from the reservoirs of Gujarat under its management, on the other, GoG in a letter to the Executive Member, NCA dated 24/04/98 responded positively to the above-stated "Alternative Proposal' of GoMP and floated the idea of establishment of a joint stock company with proportionate shares of the majority stocks initially in favour of the three states, which could be later on diluted in favour of private promoters after the project attained viability. Unfortunately, no formal agreement on this proposal could be reached by the party states, and GoMP expressed its intention to carry on fishing activity independently within its territorial jurisdiction strictly in accordance with the provisions of the NWDT Award.

3.13 From the proposals submitted by the three concerned state governments, it appears that two broad organisational forms have been debated in the past—whether the proposed policy making body called Inter-state Fisheries Board would head a cooperative structure or a joint stock company structure, which will decide on the composition as well the capital structure, among other aspects, of the proposed body. While we shall perform a SWOT analysis of these two broad organisational formats in the next section, it is important to highlight a few missing aspects, even at the cost of repetition:

- Fishing in such a vast water body cannot be done in a meaningful manner without taking care of the sustainability and livelihood issues of the fringe population—mostly PAFs and predominantly part-time or full time fishermen. How the proposed body would take care of the livelihood issues of these people with a meagre and rather uncertain fisheries yield rate, at least to start with, seems to have escaped the imagination of the proponents of fisheries in the SSP.

- In spite of concerns expressed by NCA over this matter time and again, not much thought seems to have been given to the issue of getting further clearance from the MoEF on the details of an operational plan on fisheries in the SSP, which is almost a statutory requirement.
- The emphasis has all along been on fisheries to the exclusion of other resource connected allied activities like forestry, agriculture, animal husbandry, eco-tourism etc. to tap the economies of scope.[16] This is rather surprising in view of the fact that most of the fringe population already or likely to get involved in fisheries are already pursuing some of these allied activities as supplementary sources of income, when current returns from fisheries are low and beset with multiple risks for various reasons. Probably the need for close coordination across several Governmental Departments to achieve much more broad-based cooperation to tap the economies of scope—a matter still unknown and untried in the Indian context—seems to have acted as a serious hurdle. A strong sense of autonomy and independence, probably coupled with some *de facto* sense of hierarchy, seems to have constrained the attitude and behaviour of our Departments and prevented them from achieving a global optimum (in contrast to a local optimum implying less welfare and less social good). The issue is whether we can think of an organisational format which can impart enough flexibility to it and thus help internalise the externality effects the relevant Departments are suffering from. This is a matter which we shall address in detail in the concluding section, based on several well-known institutional experiments being performed in India and abroad.

3.14 Fortunately, in course of our deliberations with the relevant officials, the concerned Departments seem to have displayed enough maturity and flexibility to negotiate for a common ground and come up with a reasonable solution to this more than two decades old

16. GoG proposal on joint stock company seems to have a better appreciation of this point, though economies of scope have not been explicitly referred to.

problem. GoG seems to appreciate the difficulty of achieving a joint stock company solution to begin with. GoMP too seems to appreciate the difficulty on the part of the other two state governments to agree to its proposal for a common Board with majority control in its favour, which practically rules out any difference of opinion even at the policy making level.[17] Although any formal change in its stand would require ratification by its Cabinet of any fresh mandate replacing the earlier stand, the overall environment seems favourable to this change. Although GoM is not currently ready for private share-holding in the proposed body—at least to begin with, it is consistent in its support to the formation of an Inter-state Fisheries Board with contributions from the three states in proportion to their geographical coverage of the water body.

Section 4: A SWOT Analysis of Major Organisational Forms

4.1 Consistent with the constitutional provision of fisheries as a state subject and joint ownership of the Sardar Sarovar Reservoir, formation of an Inter-state Fisheries Board with representations from the three party states is a logical necessity to provide a common platform, wherefrom necessary policy measures can be initiated and guided to tap the fisheries potential of this vast water body. However, there are several important issues, which need to be satisfactorily resolved in order to make this body operational and effective for the purpose in hand. The first issue is whether and to what extent this policy making body would address allied sector activities on the same resource base in order to overcome limitations of a narrow fisheries focus and at the same time to achieve a broader-based cooperation cutting across multiple government agencies. The other important issue is whether an exclusive cooperative structure or a joint stock company format should be employed to provide the ways and means for implementing the policy decisions. This section will be devoted exclusively to addressing the second issue, while leaving the first issue to be addressed in the concluding section.

4.2 Table 4.1 summarises the SWOT analysis performed on a

17. It is important to point out that in the UN General Assembly each nation—whether large or small—has an equal and single vote, so that each of them can contribute equally towards policy making.

cooperative approach with the case of fisheries development in the SSP kept in the background. The strengths of a cooperative form of organisation, especially in the backward linkages component, probably continue to be its point of attraction in the minds of most state governments in the context of fisheries development in the SSP, in spite of its multiple weaknesses and demonstrated failures in most segments in most parts of the country. The most serious question is how to capitalise these cooperatives. Even if the state governments capitalise them, in spite of their resource crunch and low and/or uncertain returns on investment in SSP fisheries, possible threats as highlighted in the SWOT analysis table would still loom large. But considering the opportunities for a cooperative approach to the nature of the problem in hand, it would not be prudent to discard this cooperative approach at least in the initial stages. So, attempts need to be made to base cooperatives on carefully developed SHGs, which will be easier to monitor using peer-pressure,[18] besides developing a broader based cooperation through integration of fisheries with allied and supplementary activities, which the PAFs and fringe population are in any case practising or are keen to practise in the near future, given the pressure of economic circumstances on them.

4.3 Although the strengths and weaknesses of joint stock companies are respectively the weaknesses and strengths of cooperatives and *vice versa*, Table 4.2, in order to make things explicit, has performed a SWOT analysis of the joint stock company approach to the problem. Given the observed strengths and opportunities of this mode of organisation, on the one hand, and the vast potential of the SSP for a diversified but integrated resource use around fisheries, on the other, it is quite natural in an era of market-based approach to look for this joint stock company approach—initially with government equity capital, but gradually moving towards raising capital from the market. Considering this futuristic view, GoMP is perfectly justified in suggesting this

18. It is a sad commentary on the growth of most cooperatives in India that they were often artificially imposed from above, rather than allowing the fundamental principle of self-help to work itself out slowly but steadily over time to give birth to cooperatives, if necessary, as higher-tier bodies, thus building up a solid foundation even to our primary level cooperatives.

Table 4.1: A Summary of the SWOT Analysis of the Cooperative Approach

Strengths	*Weaknesses*	*Opportunities*	*Threats*
1. Usually strong in backward linkages with the resource users.	1. Supply of capital as the most serious constraint due to failure of property rights on cooperative capital. The underlying problems of inadequate and risky return on investment in cooperative capital are referred to as horizon, portfolio and control problems in the literature.[19]	1. Due to community orientation, can take a holistic and integrated approach in terms of resource use and catering to multifarious needs of members.	1. Can be easily taken for a ride by vested interests like bureaucrats, politicians and even professionals/employees to the neglect of members' interests, while para-statal cooperatives dependent on government support in the form of share capital and subsidies may continue to drain the public exchequer.[20]
2. Accepts centrality of the needs of members, the resource users, and responds to the same	2. Suffers in terms of human resource development and leadership crisis affecting continuity and sustainability—an aspect often reflected in weak forward linkages (e.g., post-harvest management including marketing).	2. Can easily rope in women, children, weaker sections of people, NGOs and environmentalists to build up a broad-based stakeholder cooperation.	
	3. Very slow in decision-making due to internal democratic process.	3. Can be solidly founded on a vibrant SHG movement.	
	4. Usually very weak in networking, strategic alliances, even failing to achieve 'cooperation among cooperatives'. 5. Often fails in lobbying for itself. 6. Absence of natural acquisition and merger, and statutory binding on local orientation, prevent growth matching with natural market size		

19. *See*, for example, chapter 3 of Datta, Samar K. (Edited), *Cooperatives in Agriculture—A Millennium Study* (Volume 24), Academic Foundation, Delhi, 2003, for details.
20. GOPA Consultants (1991) made these remarks about fisheries cooperatives in the state of MP: "………most of the presently existing fishing cooperatives are 'fake' cooperatives, existing often in name only, and registered mainly in order for the members of the cooperatives to qualify for government support and subsidies" (*ibid*, p. 28).

Table 4.2: A Summary of the SWOT Analysis of the Joint Stock Company Approach

Strengths	*Weaknesses*	*Opportunities*	*Threats*
1. Can easily raise capital from the market	1. Non-starter for projects with high risks and low return—e.g. those involving poor and backward people in difficult terrains.	1. Very prompt in tapping economies of scale and economies of scope, thus often enjoying first-mover advantages.[21]	1. Craze for short-term profits may sacrifice longer-term (e.g. ecological) goals and concerns of weaker sections of people—especially the fringe population, thus a high growth company often taking the society for a ride.
2. Can easily and continuously attract beautiful minds like professional managers and entrepreneurial talents	2. Often lacks social and longer-term perspective due to over-emphasis on short-term gains in the form of profits.		2. Government corporations, too, may impose heavy losses on the society due to usual mismanagement, even if such bodies may not impose other types of losses as mentioned above.
3. Usually very strong in forward linkages including grading, processing and marketing	3. In the absence of conscious development of stakeholder cooperation, small and distant stake-holders like labourers, employees, raw material suppliers, retailers etc. are held in a purely fixed-claim vendor relationship *vis-à-vis* the owners, the suppliers of equity capital.		

21. The history of expansion of Walmart is an interesting case in point. The story of Lucky Goldstar, a South Korean multinational makes another interesting reading in application of the principle of economies of scope in its growth (for some more details, see Paul Milgroms & John Roberts, Economics, Organisation & Management, Prentice Hall Publication, 1992, p. 543).

alternative and GoG is perfectly justified in its attempts to streamline and articulate this approach in its dreams. However, two major hurdles are likely to stand in the way of realisation of this dream—at least in the initial stages. First, as pointed out in the SWOT analysis table, no one, and even the resource-poor states, is likely to invest in a company, if returns are low and risky. So, the necessary amount of capital investment may not be forthcoming unless proven records are established to attract the investors. Second, if a joint stock company becomes busy in making short-term profits to justify its investments from the very beginning, longer-term developmental and ecological goals—so crucial in the present context—may take a back seat. These limitations, however, do not imply discarding this option for all time to come. Nor do these observations imply that we should not try to reap the strengths of a corporate solution, especially in the forward linkages through roping in (i.e., through strategic alliances with) a suitable corporate body with proven records and longer-term commitment to the cause of rural development into the game. Thus, while a pure cooperative or a pure corporate strategy may not work to start with, a healthy and feasible combination of the two may work. This will in fact be the thrust of the current approach to the problem, which will be elaborated and articulated together with its theoretical underpinnings and examples in the next section.

Section 5: The Proposed Organisational Structure and the Road Map

5.1 In order to arrive at the proposed organisational structure, this concluding section has attempted to blend several carefully chosen institutional ingredients. First, it has tried to apply the principles of property rights, when applied to a vast and complex asset like the SSP, having multiple stakeholders, besides applying a negotiating framework popularly known as the Coase Theorem. These principles are used to critically examine the logical validity of the solution suggested by the GOPA Consultants in 1991 as well as the two broad and pure organisational formats—namely, a cooperative or a joint stock company approach—which have been proposed and debated in this context. Second, it has attempted to draw appropriate lessons from the functioning of the Tungabhadra Board, which is often cited as a benchmark case for fixing the functioning of the Inter-state Fisheries Board as proposed in the current context. Third, it has attempted to draw lessons from the functioning of SHGs, as for example is the case in the Indian Sunderbans, to highlight the need for tinkering with the traditional cooperative structure to evolve an integrated approach under a suitable leadership to handle the problems of very poor PAFs in a backward region. Lastly, the relevant literature on the Lake Victoria Fisheries Organisation (LVFO) is used to cull out the necessary ingredients for developing the blueprint for a suitable institutional setup in the current context.

5.2 An ideal way to assign property rights on an asset is to ensure that there is a perfect match between the residual controller and residual claimant to the asset, which is usually the case for simple private property resources (e.g., a car). A residual controller would have the ultimate authority to fix rights of use to the asset under consideration. Even if he delegates certain rights of use to a hired agent, he is at liberty, of course subject to the laws of the land, to alter or fine-tune those rights. A residual claimant enjoys all benefits and returns from the asset he owns after meeting any fixed claims of hired agents, who may be permitted to use the asset under an agreement with the owner. Our legal systems are designed so as to protect both residual control and residual claim rights of the owner of an asset, while balancing the interests of possible hired agents of the owner *vis-à-vis* the same for the owner.

5.3 Unfortunately, most assets are complex types with multiple stakeholders, as in the case of the SSP. Hence, application of the above-stated principles becomes a nearly impossible proposition. In this context the famous Coase Theorem comes to the rescue by setting up a process through which the multiple stakeholders can negotiate among themselves to fix their rights and obligations with respect to the asset in question. Three important conditions are laid down by the Coase Theorem to ensure that the negotiation process ensures generation of maximum value through appropriate utilisation of the asset. First, the negotiation process must start from an initial unambiguous assignment of rights to the asset in favour of one stakeholder or the other, so that the negotiators do know which right to trade (i.e., negotiate) and with whom. In case of the SSP, the NWDT Award has set up the initial rights allocation, which the party states can always fall back upon. Second, for reallocation of rights on the asset to improve upon performance (i.e., better results through use of the asset), a platform is needed wherein the concerned stakeholders to the asset can negotiate with one another at minimal transaction cost. It is precisely because of the transaction cost that the current exercise is looking for a suitable institutional setup, wherein the stakeholders to the SSP can easily negotiate with one another. Hence, representation of all important stakeholders to the asset is a must; else externalities—whether positive or negative—will arise from the stakeholder left out of the platform. That is, an external stakeholder, if left out, can get away with certain benefits without incurring any costs (in case of positive externality), thus creating a leakage. Alternatively, a left out external stakeholder may suffer negative externality in the hands of the internal stakeholders, thus increasing the likelihood of creation of nuisance value by the former and for the latter group of stakeholders. Thus, automatic internalisation of externality through roping in all important stakeholders to the asset under a suitable umbrella organisation is the second guiding principle of the Coase process. Third, the negotiation process has to be guided by rules in every stage such that no one stakeholder can coerce other stakeholders into accepting a deal best suited to him. In other words, during the negotiation process (or the policy making exercise, as in the current context), no stakeholder can take the others for a ride (thus, generating what is called 'wealth effect' in its favour). Experience suggests that this Coasian negotiation process is best adhered to if the most intimate stakeholder (meaning the side, which would suffer the most under an improper use or underutilisation of the asset) is put at the centre stage of the negotiation process. The identity and stake of the most intimate stakeholder may however vary over time and space. Given the unorganised character of the PAFs and the fringe population to the SSP at the initial stage, the Fisheries Departments of the concerned states may be deemed to be the most intimate stakeholders. But as these people improve their lot and get more and more organised, they may replace the states as the most intimate stakeholder and negotiate for a fixed contractual (lease) payment to the states. Alternatively, if the fringe population gets absorbed and economically settled in diversified activities around the SSP, only some of them may like to continue to work as hired fishermen at fixed contracts under a joint stock company—whether run by the states or even by a private entrepreneur. The purpose of the current exercise is to simulate the dynamics of the situation under reasonable assumptions and using the conceptual framework as spelt out in this section.

5.4 Table 5.1 summarises the institutional setup as proposed by the GOPA Consultants (1991), which have beautifully conceptualised the problem by dividing the tasks into three components—regulation, enforcement and commercial exploitation, besides assigning responsibilities to appropriate bodies to perform these tasks. They have also proposed suitable investments in fisheries infrastructure to conduct fishing in the SSP (their suggestions reproduced in Box 5.1). While there is no disagreement over the needed investments in infrastructure and the broad tasks to be performed, we find some gaps in the logical structure of their proposed institutional setup, which are highlighted below for necessary remedial actions during the current exercise:

Table 5.1: Institutional Networking as Proposed by GOPA Consultants (1991)

Tasks	*Institutional arrangement*	*Details*
1. Fishing regulation	Inter-state Board	Composition: Representative of GoI; Director, Fisheries, MP; Director, Fisheries, Maharashtra; Director, Fisheries, Gujarat.
2. Enforcement	Line agencies of state governments	Functions: Issue of licenses Control of fishing activities.
3. Commercial exploitation	Three-tier cooperative set-up	Functions: Primary societies (fishing) Two federations (marketing, supply of ice, gear etc.) Cooperative Apex Body (reservoir management, stocking, hatchery etc.).

- The fact that GOPA Consultants have assigned the Chairmanship of the proposed Board to NCA with the responsibility of coordinating the fisheries development in the SSP, in spite of some valid reasons put forward by them,

Box 5.1: Investments in Supporting Infrastructure as proposed by GOPA Consultants

It is proposed to build landing centres as follows:
1. in Gujarat (dam site),
2. in Maharashtra (ferry site, tentative location, possibly also on the Gujarat side of the reservoir, depending on development of road/railway infrastructure by 1997, first fishing year)
3. in Madhya Pradesh (mud road access point, and Barwani).

Each landing centre would have the following features:

Sites:	4, yet to be selected
Capacity:	2.5 tons of fish/day

Ice plants would be established as follows (one at each extremity of the reservoir)

Sites:	2, yet to be selected
Capacity:	10 tons of ice

....A margin has been provided and each site equipped with a 10-ton ice plant. The cost increase compared to a 5-ton ice-plant is not considerable. Daily peaks are taken care of by the storage capacity of cold stores.
Cold storage units would be set up as follows (one at each extremity of the reservoir next to the proposed ice-plant:

Sites	3, yet to be selected, one in each state
Storage capacity	20 tonnes of ice

For enforcement of fishing regulations lake transport has to be provided.........

goes against the principle of ensuring perfect match between residual control and residual claim rights. While only the party states would have the residual claims, the residual control rights would be tilted in favour of NCA, which is a mere coordinating and monitoring body of the Government of India with no ownership right on fisheries in the SSP, nor any technical expertise of its own to undertake fisheries. Moreover, this apportionment of fisheries property rights and obligations in favour of the NCA may not only overburden the NCA, but also create a possible misunderstanding between the states and the centre in view of the constitutional provision for fisheries as essentially a state subject.

- Probably the GOPA Consultants were justified at that time in expecting World Bank funding, besides NABARD's support for fisheries development and especially for infrastructure development. But now that World Bank funding is not there, nor is NABARD very keen to fund fisheries projects of this type with risky outcome, the question of funding support looms large. Given the capabilities of fishermen's cooperatives and the state governments' financial status, on the one hand, and at best limited returns from fisheries alone, on the other, probably it would be prudent to minimise massive investment expenditure to begin with.

- Given the socio-economic status of the PAFs and fringe population and the SSP's moral commitment to the ousted households, fisheries alone may not be a sustainable proposition. Given on the one hand resource connectivity across uses, and on the other hand the fringe population's demonstrated eagerness to undertake several allied activities as supplement to one another, the GOPA Consultants seem to have undermined the possibility of an integrated resource use around fisheries in the SSP, so that even when fisheries alone may not be an attractive commercial proposition to begin with, a set of allied activities together can reinforce mutual viability. An integrated approach is a must considering the fact that any organised fisheries activity in the SSP requires prior intimation to the MoEF. And if the fringe population are indeed willing to practise fisheries and forestry together for meeting their needs, there is no reason why some of these families cannot be organised into joint forest/ecology management groups (especially, in the form of SHGs) and promoted alongside fisheries. This approach will also create a space for foresters and environmentalists going alongside the fishery officials with better coordination. When ecological effects of fishery or ecological effects of other activities on fisheries cannot be totally ruled out, it is always pragmatic to internalise the possible sources of externality by also seeking representations from the Departments of Environment and Forests into the proposed Board.
- Village Ponds Development may remain a separate project, but it cannot be totally de-linked from the fisheries development in the SSP, as success of the former would have considerable beneficial effect on the viability and sustainability of SSP fisheries.

5.5 While attempting to draw necessary lessons from the

functioning of the Tungabhadra Board for Sardar Sarovar, it may be pointed out that the former is a comprehensive body in charge of undertaking all kinds of activities including irrigation, power and fisheries simultaneously, unlike the SSP, where fisheries activity is apparently de-linked from others. Fisheries activity is being performed by a separate Fisheries Wing under the guidance of a committee, which includes some fisheries personnel as members. This Wing is also looking after net-making, ice plant and other units. Absence of detailed data prevents separate evaluation the performance of fisheries in this reservoir. However, the fact that cage and pen culture has been introduced in this reservoir, apparently with a fair amount of success, raises hope about application of the same technology in the context of the SSP. From the available documents it is not clear whether the objective of socio-economic development of the PAFs, which is a clear commitment on the part of the relevant state governments in case of the SSP, is there in this project and if so, how it is being fulfilled. However, the task of fishing in the reservoir has been auctioned out to a private party in the form of a leasing contract—a practice, which can be applied with suitable modification to preserve the rights of the fishing population in the fringe areas of the SSP, if the proposed fisheries body doesn't want to get involved in costly and risky investments in harvesting and post-harvest management—at least in the beginning.

5.6 There are important lessons one can also derive from the functioning of SHGs in this country, as there is a strong similarity of the socio-economic conditions of the SHG population in most of the areas with those of the fringe population in the SSP, though the latter have some advantages in terms of their entitlement to special packages of government assistance due to eviction from their traditional places of living. Empowerment through training and awareness creation, careful development of institutions on the solid foundations of SHGs, and handholding role of one local NGO under a barefoot leadership seem to have unleashed a huge potent force for development of these people around fisheries and allied activities, using the various existing rural development schemes of different Departments of the Government of India.[22] Given the generally poor performance of

22. For details, see Datta, Samar K. *et al.* (forthcoming), Developing India's Strategies to respond to the Global Debate over Fisheries Subsidies, Allied Publishers, Delhi, Chapter 4.

cooperatives in most parts of the country and in most segments, this experience of building up of cooperatives through the SHG route provides a novel way of rebuilding the country's confidence in cooperatives. The existing fisheries cooperatives in this region may be re-engineered to become integrated fisheries cooperatives of SHGs, by organising the cooperative members into SHGs with diversified set of activities, as in the case of Sandeshkhali-II. It appears this task can be better performed with fresh members from this relatively unexplored region. A strong backing of the political as well as administrative system together with a strong commitment of local NGOs is a crucial factor for initiating this process. Exposure trips and training of carefully selected local leaders and officials to well-known successful experiences of this kind in several parts of the country can unleash this force. The organisational format proposed in this study can provide only the logical structure or institutional mechanism to realise the goal.

5.7 We now turn to the example of the Lake Victoria Fisheries Organisation (LVFO), which has important lessons to provide in the context of the SSP. Except for the fact that Lake Victoria is an inter-country natural reservoir spread over three sovereign states (thus involving support from the UN bodies, especially FAO), whereas the SSP is only an inter-state body artificially created within one sovereign country and thus involving the additional responsibility toward the fringe population—mostly PAFs, the problems are very similar. So, with appropriate modification (and possibly simplification), the institutional design already suggested for LVFO can be adapted to suit the needs of the current problem—namely, fisheries development in the SSP.

5.8 Three types of documents on LVFO are found useful in this context—first, the Conventions of the LVFO with various articles of association across the three sovereign states as well as the details of the organs of this body and their functioning; second, the strategic programmes of the organisation covering its multifarious mission statements; and third, the model for community based fisheries system (as prevalent in one of the three states, Tanzania), detailing the interrelationship and division of tasks between the Fisheries Division of the government and the community level fisheries cooperatives. We propose to undertake appropriate revisions and simplifications of these materials to cover the following aspects:

1. Considering the multifarious activities that are actually being undertaken by the fringe population under force of circumstances to tap the economies of scope arising out of the same resource base, we are in favour of being absolutely explicit on one point—that we would like to prescribe an Inter-state Board for Integrated Fisheries Management in the SSP, based on a broader based cooperation across some crucial Line Departments stakeholders to allay the fears of possible ecological degradation of the reservoir's environment.
2. Considering that the success record of cooperatives in this country has so far been poor, especially in raising capital (both human and non-human) and in making forward linkages, we are favour basing cooperatives on a strong SHG movement in the backward linkage component. As is being practise by GoMP, this will be done by offering exclusive rights of fishing under a regulated system only to the local fishermen and their bodies, while leaving the forward linkage component to be made under a longer-term lease contract a single well-known business house with a proven records of long-term commitment to the cause of rural development alongside business promotion. In this way, we propose to minimise the risks of massive investment by the state governments and the fishermen as well as allied resource users within the fringe population. The lease money from fisheries contract supplemented by grants and support from various Line Departments (especially to promote SHGs) can be used by the Inter-state Board to finance diversified rural development activities for the fringe population not only to facilitate development of this population, but also to initiate the groundwork for moving towards a joint stock company format—whether by the states or by a private company under regulation or by the cooperatives of integrated fish farmers (if they can come up to that level)—to tap the potentials of integrated resource use around fisheries on this reservoir. If these tasks are mostly performed by appropriate Line Departments with a minimal manpower support of the Inter-state Board, a part of the lease revenue can be saved and

distributed to the party states in proportion to their geographic title to this water body. If some costs are to be incurred prior to revenue generation, these costs too can be shared in the same proportion.

3. Periodic (preferably quarterly) evaluation of the performance of fishing and rural development activities towards the community's socio-economic development will decide on the future course of actions. All we can suggest is an indicated roadmap to further glorify the goals of the proposed organisation.

5.9 Based on the discussions and findings in preceding paragraphs, we present below the salient features and highlights of the proposed 'Inter-State Board for Management of Integrated Fisheries in the Sardar Sarovar Reservoir' (IBM-IFSSR), which if agreed upon by the concerned party states, may be established through an MOU across these states:

- The broad objectives of this organisation would be to foster cooperation among the Contracting Parties (i.e., the states of MP, Maharashtra and Gujarat) to harmonise international, national and state-level measures for sustainable utilisation of the living resources of the reservoir and to develop and adopt conservation and management measures for promotion of fisheries and related activities.
- The above-stated broad objectives must be accompanied by several mission statements, as in the case of LVFO, which will guide the functioning of this body:
  - A Healthy Sardar Sarovar Ecosystem Vision Statement
  - Institutional/Stakeholder Partnerships Vision Statement
  - Integrated Fisheries Management Vision Statement
  - Coordinated Research Programmes Vision Statement
  - *Information Generation, Flow and Exchange Vision Statement*
- The specific objectives of IBM-IFSSR would be:
  - Proper management and utilisation of fisheries and other related resources;
  - Capacity building of existing institutions, and development of additional institutions;

- Acting as forum for discussion on impact of initiatives, with strong liaison with existing bodies and programmes;
- Conducting of research;
- Undertaking training and extension;
- Introduction, monitoring, and control of suitable animals and plants;
- Clearing house and data bank for information;
- Adopting budgets, seeking funds, formulating plans to support activities of contracting parties; and
- Any other function necessary in connection to the above.

- The Board, as in the case of LVFO, will have the following organs:
  - A Policy Steering Committee, consisting the Principal Secretaries/Secretaries in charge of Fisheries in the three party states, Principal Secretaries/Secretaries in charge of Environment and Forests in the same three party states,[23] Member Secretary of NCA, a national level management expert, and a specially recruited professional as the Executive Director of the Board, who will act as Member-Secretary for this Committee. This Committee will take only unanimous decisions in the presence of all committee members or their suitable representatives. It must meet half-yearly at a convenient venue. Secretary of Fisheries of one of the three member states will occupy the position of Chairman of this Committee on a yearly rotational basis in an alphabetical order corresponding to the names of the states.
  - An Executive Committee, consisting of the technical heads (usually Director/Commissioner) of the three concerned State Departments of Fisheries, Director (Environment) of NCA, one subject matter-specialist as

23. The justification of involving the Departments of Environment and Forests lies in the principle of internalisation of externality, to facilitate environmental clearance of the activity streams through in-house consultancy with the relevant Department, on the one hand, and also to promote SHGs/cooperatives of SHGs to nourish eco-development activities/services in the spirit of joint forest management alongside the SHGs/cooperatives of SHGs for fishing.

an outside expert, and the Executive Director of the Board. This Committee must meet every quarter to initiate and evaluate action plans, which must be approved by the Policy Steering Committee. As in the case of the preceding Committee, this Committee too will take only unanimous decisions in the presence of all committee members or their suitable representatives. Director/Commissioner of Fisheries of one of the three member states will occupy the position of Chairman of this Committee on a yearly rotational basis in an alphabetical order corresponding to the names of the states. However, to ensure better understanding, confidence-building and coordination across the three member states, the Chairpersons of these committees should not be from the same state at the same time. A representative of the NCA will be on both Committees to facilitate resolution of possible conflicts, if any, across the three states.[24]

– The Executive Director of the Board should have an office, with a small set-up as far as possible, to coordinate the functions of the Board with the relevant Line Departments in the states and at the centre. He must be a professional person with a proven record, ability and vision to manage the organisation as well as provide the necessary leadership to achieve the vision statements of this organisation. He must be on a term appointment initially for 3-5 years, which can be extended further based on his performance. He becomes the link between the two Committees as stated above, besides being the liaison person between the Board and the Company in charge of harvesting and marketing of fish and other allied sector products (e.g., medicinal plants and herbs from the embankment and fringe areas on both sides of the reservoir) and services (e.g., transport, eco-tourism)

24. These two Committees may co-opt other related Line Departments and/or organisations (like funding agencies) as and when required as invitees without voting power.

from the reservoir, on the one hand, and the different tiers of cooperatives of the SHGs formed by the fringe population, on the other. He will receive and examine various proposals of development from these bodies and recommend them for approval by the Policy Steering Committee. As in the case of LVFO, the Executive Committee would have the power to form various sub-committees (with support from the Policy Steering Committee) to develop various business development plans and commission/undertake researches, as well as to evaluate the functioning of various organs of this organisation to generate necessary policy inputs for the Policy Committee.

- There should be a single harvesting and marketing contract for the entire stretch of the reservoir, wherein a well-known national company with a proven record and longer-term commitments to rural development must be roped in. The Board should enter into an MOU with this company—at least a medium term one of five years, wherein the Board may provide some incentives to the Company by undertaking some investment in infrastructure (like storage facilities, landing centres etc.), for example, under NABARD's RIDF scheme. If found necessary, the Board should enter into strategic alliances or networking arrangements with several companies simultaneously, as has been the case with successful business concerns like the Gujarat Cooperative Milk Marketing Federation (GCMMF).
- As in the case of LVFO, fishing and allied activities in and around the reservoir can be undertaken as a matter of exclusive right to be granted only by the PAFs and fringe population against a token licence fee, so that the Board would have an instrument in hand to control access to the various facilities of the reservoir and handle possible run on the resources. The fishermen would be paid fixed catching fee, (as periodically decided by the Board), by the contacting company. The fringe population need to be initially organised into SHGs

engaged in various feasible activities of their choice. Existing schemes of the GoI can be tapped to promote these SHGs with some financial stakes, unlike in cases of most cooperatives wherein the members have very little financial stake equal to the value of a share. Promotion of SHGs must be undertaken by spearhead teams on turnkey basis, as was done by NDDB in the initial years to promote village milk cooperatives. The SHGs should be encouraged to organise themselves into village or Panchayat level cooperatives. Depending on the progress of activities and interests of the primary level cooperatives and the economics thereof, these primary level cooperatives ought to be encouraged to form higher-tier bodies, as in the case of the AMUL model. No higher tier should to be formed if it cannot perform a net value-added function, which the lower tier cannot do profitably. In other words, the functions of each tier must have a strong economic logic in economies of scale, scope and value-addition. Since development of the backward linkages through SHGs and cooperatives is a long-drawn out process, as in the case of AMUL, we suggest beginning the process of harvesting and marketing fish with the help of a contracting company. Regular schemes of various Line Departments should be used to organise the fringe population and the PAFs into SHGs and higher tier cooperatives of the SHGs. The Board's job will be to merely coordinate and spearhead this process. As in the case of the Tanzanian community fisheries, the responsibilities of the SHGs/ their cooperatives engaged in integrated fisheries must be clearly demarcated *vis-à-vis* the same for the Fisheries Departments and the Fisheries Division of the Board.[25]

5.10 As highlighted in the earlier paragraphs, there is immense potential for growth in fish productivity and diversification of

25. There will be need for a separate Fisheries Division only after other allied activities gain in importance.

activities. However, all these things cannot be achieved overnight in a single stroke. We have therefore the following notes of caution or advice on the future road map of the proposed organisation:

- A joint stock company structure under the Board doesn't seem to be a feasible proposition to start with. But given sufficient hard work by a visionary leadership of the proposed Board, it can be achieved in the longer run, through gradual introduction of scale, scope and value-addition activities with the help of a contracting company to perform the forward linkage activities beginning from harvesting of fish.
- In the initial phase, the emphasis should be on augmenting fish productivity of the reservoir through application of various institutional and technological innovations, as indicated earlier. The only diversified activity, which ought to be promoted from the beginning, is promotion of forests and ecology with the help of PAFs and interested fringe population in the spirit of joint forest management, so that the cause and concern of the MoEF can be adequately taken care of and their support and encouragement can be ensured from the very beginning. Other diversified activities can be taken up in full swing in successive stages, only after the first two primary activities are well established.
- We do not recommend heavy investments in fisheries, ecology, infrastructure development or marketing activities either by the states or by the SHGs or their cooperatives to start with, as we feel that, given the current state of affairs, no one will be willing to risk massive investment. For this reason have recommended a 'mongrel' organisational format to start with—a refined cooperative structure in the backward linkages (i.e. at the bottom) and a company structure on the top (i.e. in forward linkages), but on fixed contract basis to avoid risks.
- The gradual unfolding of the organisation will depend on the relative capacity building and demonstrated efficiency of the three major stakeholders—the three partner state governments forming the Board, the Company handling harvesting, storage, logistics and marketing, and the fringe

population inclusive of the PAFs. We can contemplate three possible outcomes: (i) the SHGs and their cooperatives replicate the AMUL success story, so that their members take charge of everything (i.e. all residual risk) by calling the shots and relegating the other two partners as mere fixed claimants (of lease rent and fixed commission charges); (ii) the three state governments together under this Board develop sufficient understanding, mutual trust and confidence to run the organisation as a public limited company with full or majority shares held by them, in which case the Board becomes the residual claimant, while the other two partners will be relegated to the status of fixed claim agents or vendors; and (iii) the private company gains in experience and confidence to undertake more and more investments in fishing and related activities so as to enhance profitability of integrated fisheries on the reservoir and thus be in a position to pay progressively higher and higher lease rents to the Board and also progressively higher and higher wages to the fringe population for their services.

A convex combination of these three options may also emerge and persist for a long time, if not for all time to come, as implicit in the proposal of GoG. Only the future can tell which course history will take, but if adopted earnestly and with vision, any of these broad courses will do good to the cause of the society and the stakeholders to the SSP.

Section 6: Concluding Remarks

6.1 Although the suggested solution raised a lot of hopes and enthusiasm among the three major stakeholders amidst an apparent agreement amongst them for a coordinated solution, it is learnt that slowly but steadily they started developing cold feet as they found that the proposed Inter-State Board would vest substantial control in NCA and may thus interfere with the autonomy of individual states as granted by Indian Constitution in pursuing independent fisheries policy. So, all the three states, as it is learnt, gave up the idea of having a coordinated body like Inter-state Fisheries Board, in spite of its potential benefits as elaborated in the preceding section, to fall back upon the old option of pursuing independent fisheries activities within their own geographic territories within the reservoir (as per the NWDT Award). Going back to *status quo* at the end of a prolonged period of negotiation and several rounds of expert consultation may look like falling back upon a second-best or even third-best solution, but it has a rationale, which probably needs to be brought forth and understood in the interest of possible future negotiations on utilisation of inter-state assets or solution of inter-state problems of similar types.

6.2 This is where the celebrated Coase Theorem[26], proposing conditions for achievement of value maximising property rights of assets through negotiations, comes in as a handy tool in explaining failure of the theorem. Coase has prescribed three fundamental conditions for success of negotiations leading to a value maximising outcome. First, the initial property rights must be clearly defined to create an understanding of what each stakeholder is gaining or losing from the *status quo*, so that the necessary urge is created for success in negotiations. Second, the transaction costs of switching over to any new regime of property rights must not be so high as to make the gains from negotiation over this fresh regime of property rights an unattractive proposition. Third, for the process of negotiation to continue in a free, fair and competitive spirit, no stakeholder should be able to exert 'wealth effect' or extraordinary bullying power to scuttle the negotiation process. From an understanding of what has happened of late on the fishing rights issue on this vast water body, it appears *prima facie* that the Coase conditions have failed to assert the value maximising solution. Naturally, a lot more research and probing are necessary to bring out the details, which probably can provide clues for success in future negotiations.

Acknowledgements: The author would like to express his gratitude to Sardar Sarovar Narmada Nigam Limited, Govt. of Gujarat for funding the base study, and also to Ministry of Animal Husbandry & Fisheries and Narmada Control Authority of Government of India, besides all the three involved states of Madhya Pradesh, Maharashtra and Gujarat for sharing valuable information and ideas. The present paper is based on the base study report submitted earlier to the funding agency.

---

26. See, for example, Milgrom & Roberts (1992) for details.

Appendix 1
Story of SSP Ousted Population Now Rehabilitated in Suka Village

In order to understand the stakes and concerns of the local population in fisheries in Sardar Sarovar, we visited a village called Suka, about 15-20 km from the dam site, and conducted a participatory rural appraisal (PRA) exercise. This village comprises about 200 families, of which about 150 were relocated in 1985 from their original village, Panchmauli, now submerged by the reservoir. The remaining 50 families—another tribal group different than the one from Panchmauli—are traditional inhabitants of Suka, who sold part of their land to the government for resettling the newcomers.

The PAFs were each given 5 acres of land, besides a house with electricity connection, in Suka to pursue agriculture, though with growth of population the allotted land fell short of their need for making a living. The lack of irrigation so far in this part of the command area has further restricted the possibility of a second crop beyond the rainfed crops they have been growing—a coarse variety of paddy, corn, *jowar, tur*—mainly for domestic consumption. Some of the family members had to inevitably look for alternatives—wage labour and even temporary migration to nearby cities, especially Surat, to work on construction sites. Almost all the families have fishing skills, and some of them—about one person from each family—engage in fishing, formally at the dykes as members of a fisheries cooperative and informally at the reservoir. Before shifting to Suka, they lived mostly on forest produce and partly on shifting cultivation and fishing in the river. Thus, they are still adjusting to a new occupational pattern consistent with their current resource endowment. In order to supplement their meagre income, each family now maintains two heads of cattle (cows and/or buffalo) and sell the surplus milk in the nearby official colonies.

Thanks to the efforts of the Sardar Sarovar Nigam and other line Departments, these families enjoy fairly good infrastructure in this village. The village has a school up to 7th standard. It has 15 boys and 15 girls who are matriculates and another six boys who are graduates. Regular medical services are provided by a mobile medical van which visits twice a week. There is a drinking water scheme. Extension officials of the Rehabilitation Department visit once a week, but the villagers allege that their interaction these days hardly produces any fresh results. Nevertheless two people who retired from the army have bought tractors, which they rent to the villagers at Rs. 200-250 per hour.

The villagers travel about 15-20 km on bikes to fish at the dykes and the reservoir, as mentioned above. They use gill nets which they set up in the evening and collect their catch the next morning. The Gujarat Government has provided 15 boats to the society, so sometimes a group of 2-3 people goes fishing in a boat. The minimum catch per boat they report is 10-15 kg a day, which reaches a maximum of 100 kg/day for about 15 days preceding the rainy season. They generally bring their catch to the society which arranges sale through listed traders at a rate of Rs. 25/kg. But they allege that they are yet to receive their dues from the cooperative and its traders. If the catch is good, they prefer to go to Surat to sell it, but they are often constrained by lack of suitable transport facilities. After two villagers lost their lives to crocodiles at the dykes (which they say are full of crocodiles) they stopped fishing there. Lacking appropriate equipment, they are scared to fish in the reservoir though the fishing potential there is quite high according to them.

The Fisheries Department of the Government of Gujarat claims to have stocked about 442 lakh fingerlings in the reservoir since 1995-96, whereas their stocking rate in the dykes is around 15-16 lakh fingerlings. Yield rate for the dykes, as reported by the Department, not only very low but also very varaible. It was only 4.84 kg/ha in 2002-03, but rose to 10.64 kg/ha in the very next year. Test fishing results confirm the presence of not only more than 80 per cent of the Indian Major Carps (IMC), but also about 0.5 per cent of Mahseer.

While the fisheries cooperative society is hardly functioning, the people are seriously constrained due to near-non-availability of institutional credit. Local moneylenders charge as much as 3-5 per cent interest per month and that too for small loans!

30

Challenges Involved in Operation of SSP Canal Conveyance System and Use of State-of-the-Art Technology

M.B. Joshi and *K.D. Acharya*

Canal Operation—Key to the Improved Performance of an Irrigation System

World-wide the performance of an irrigation system is measured in terms of Adequacy, Dependability, Equity and Efficiency. Sardar Sarovar Project (SSP) belongs to a few of the well studied multipurpose water resources projects in which these criteria have been given due weightage right from its conceptualisation stage and also during the planning, designing and execution. It is in this backdrop that operational aspects of SSP canal network become increasingly important to help achieve the desired level of performance. This paper is an attempt to present the uniqueness of the functional requirements of this canal network, the need for use of state of the art technology and the efforts being made in this direction.

Distinct Features of Sardar Sarovar (Narmada) Project and Canal Network

SSP will be one of the largest multi-purpose river valley projects of India. It envisages construction of a large concrete gravity dam, hydro-power complex of 1450 MW from river-bed and canal head hydropower plants and a very large network of canals and drainage channels as required to irrigate about 1,845,000 ha of land out of 3,428,000 ha of gross command area. The canal water released from the Sardar Sarovar reservoir will pass through the 250 MW canal head power house located below the Vadgam saddle dam on the right rim of the reservoir near Sardar Sarovar Dam. The water will then flow through four secondary ponds created by a series of rockfill dams, which are interconnected by open channels. The reservoir extends on the northern side from the dam site up to Gujarat-Rajasthan State border.

The Narmada Main Canal (NMC), one of the largest lined irrigation canals in the world, off-takes from Pond No. 4 downstream of the dam. It will carry a design discharge of 1,133 cumecs with a dominant discharge of 934 cumecs which would taper down to 71 cumecs at the tail end to the neighbouring state of Rajasthan in a distance of about 460 km. NMC would annually convey 11.1 billion cubic metres of water (including the share of Rajasthan). There will be 38 branches off-taking from it with a discharge varying from 320 cumecs to 3 cumecs. In SSP it is planned to construct a network of canals about 74,626 km long. The details of the network are given in Table 1.

Table 1: Canal Network of SSP Command Area

Sr. No.	*Name of Canal*	*Projected Length of canal (in km)*
1.	Main Canal	458
2.	Branch Canals	2,585
3.	Distributories	5,112
4.	Minors	18,413
5.	Sub-minors	48,058
	Total	**74,626**

This canal network is further classified into Conveyance Network and Distribution Network. The conveyance network has many distinct features making its operation a real challenge. The farthest point in the command area lies at a distance of nearly 700 km from the reservoir. A rough estimate is that it will take about 7 days for a change made at the head to get transmitted down to a tail, located at farthest point. To overcome this and to improve the response time, it has been planned to operate the conveyance system with Controlled Volume

Concept. Efficient water use is also imperative to irrigate a vast culturable command area of over 1.8 million hectares as against the availability of hardly about 9,800 MCM of surface water from the project, restricting the Delta to just 53 cm (21 inches) at the head regulator of the Main Canal.

Apart from its unprecedented size, its technical complexity also makes its operation special. Amongst a few such noteworthy features is the Saurashtra Branch Canal (SBC), which is the largest branch canal off-taking from NMC at Ch 263.20 km near Karannagar, Taluka Kadi, District Mehsana of North Gujarat. SBC is a ridge canal. Its total length 104.406 km. and it tails into Bhogavo-II reservoir near the city of Surendranagar. Along its alignment this canal negotiates the natural saucer shaped topography, involving falls (vertical drop) at three locations aggregating to 35 m drop in the initial reach and then a lift of total 71 m on the other side. At the fall locations, it is planned to install Small Hydro Power stations of 16 MW at each.

Five pumping stations are constructed in a series on SBC to irrigate 500,000 ha command area in the drought prone region. The lifting height and pump configurations at different locations are given in Table 2.

Table 2: Average Head for Lifting and Pump Configuration

Chainage in km	*Average Head for Lifting in metres*	*Pump Configuration (Cumecs x No.)*	
		Concrete Volute Pumps	*Vertical Turbine Pumps*
69.000	11.200	20 × 10	5 × 6
78.000	15.900	20 × 5	5 × 4
90.000	15.800	20 × 5	5 × 4
93.000	12.150	20 × 3	5 × 4
102.000	17.400	20 × 3	5 × 4

Functional Requirements of Canal Operation in SSP

The operation of the SSP canal system is basically required to serve following objectives:

- Required quantity of water should be delivered at appropriate time and place to the user on demand,

- The System should be capable of responding quickly to any change in demand or to any emergency condition,
- It should match the supply to the demand and maintain balanced canal operation without wastage of water, and
- The canal should operate within allowable constraints like freeboard and drawdown even under transient flow conditions.

Safe and efficient operation of this canal network has a major role to play in deciding the performance of the project in terms of fulfilling the drinking and domestic, irrigation and industrial requirements.

Safety Aspects—Freeboard and Drawdown Criteria in SSP

Irrigation demands dominate the canal flow schedules and therefore the operation has to match the variations in demand with due recognition of supply. Unsteady state flow conditions arising in the canal due to operation of the control structures need to be carefully looked into to avoid untoward incidents like overtopping of canal or failure of embankment or lining. For NMC, considerations of the safety of canal lining have dictated the following drawdown criteria:

Time step	*Cumulative drawdown*
For the first one hour or part thereof	15 cm
One hour to two hours	30 cm
Two to four hours	30 cm
Four to twenty four hours	65 cm

Other technical reasons also make system operation critical. SBC, for example, involves pumping of approximately 285 cumecs discharge to a height of more than 70 m. In spite of every possible redundancy in pumping system, its failure will cause an instantaneous rejection of such a huge discharge, which will be required to be accommodated within the conveyance system. Similarly, Kuchchh Branch Canal (KBC) also involves pumping of approximately 76 cumecs discharge to about 45 m height in two stages. The enormous time required to fill or empty the canals (while respecting the freeboard and drawdown criteria) also needs due consideration.

Multiple Uses of Water Conveyed: Domestic Needs Critical

In addition to irrigation requirements, the SSP conveyance network has to cater to the municipal (domestic) demands for 9,633 villages and 131 urban centres and industrial water demands also. Such a multiple-use canal system obviously gets subjected to frequent variations in flow requirements and unabated 24×7 operation becomes imperative. Operation of the conveyance system becomes equally critical due to the vastness of the command area, which warrants balancing of the demand and supply and no single solution is applicable to all the 13 different agro-climatic zones (Figure 1). Also, due to possible heavy downpour in any corner of the command area, there may be instantaneous rejection of water demand, leading to an emergency situation that may endanger the canal. Similarly, emergency conditions arising in any canal water pool may necessitate sudden closure/drastic reduction of discharge.

Challenges Involved in Operation of SSP and Inter-State Commitments

In the Sardar Sarovar Project, efficient operation of the canal

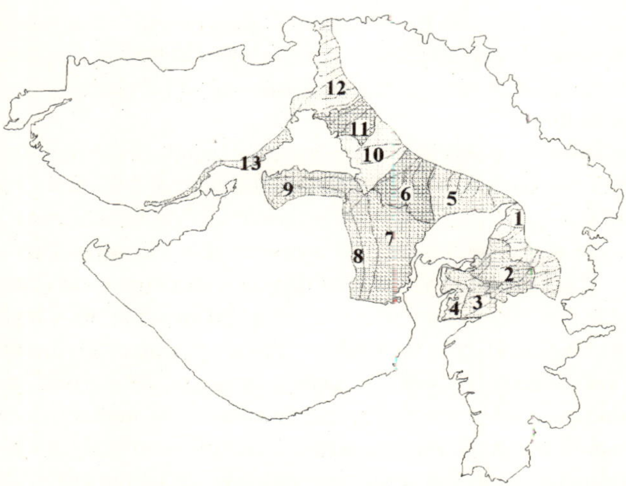

Figure 1: Agro-Climatic Zones in SSP Canal Network

conveyance system is a must in a view of fact that 75 per cent of the command area is drought prone. Low 'carry over' storage available in the reservoir and dependence on regulated releases from u/s reservoirs make such operation more imperative. Because of very limited Delta of 21´´ (53 cm) at the head regulator only protective irrigation would be aimed at and hence dependability of the irrigation system and response time becomes far more important. Operation of the SSP conveyance system has other objectives in addition to the equitable distribution of water within the agro-climatic zone. For example, as an inter-state project, it has the obligation to deliver Rajsthan's allocated share at the designed head (head-loss becomes a critical constraint in canal operation). At the same time, water supplied to Rajasthan on any date in excess of 10 per cent over its indent shall reckon against use by Gujarat (Clause IX [xii], NWDT Award). In fact, even 1 per cent variation in the discharge in the head reach of NMC is equivalent to more than 10 per cent of the annual share of Rajasthan. Rainfall directly entering the canal prism cannot be ignored in the present case, as total surface area of the NMC alone is more than 38 million m^2. Thus, the stochastic nature of both supply and demand imposes severe constraints on the system operation.

Conventional Method of Operation and its Limitations

Most of the canal systems in the country are operated in a manner known as "Conventional Operation" consisting of Scheduled delivery, Upstream operational concept, and Constant downstream depth operational method.

The limitation of conventionally operated canals is the inevitable discrepancies between the predicted and the actual deliveries both spatially and temporally. In addition there will always be inaccuracies in the flow through checks and the amount of water stored in the canal pools. The present day needs of timely water supply at critical stages of various crops and need to respond quickly to varying demands put several limitations on the conventional designs developed for relatively small irrigation projects. Since the canal system is not operated to react to actual demands, any such error will thus be accumulated at the far downstream end of canal system which would result in a typical "tail-enders' syndrome" in which tail-enders suffer from either too much or too little water. To prevent shortages of water to the tail-enders, excess water has to be supplied from the head-works. Most of the time this excess release ends up being wasted at the downstream end of the system.

Controlled Volume Concept (CVC)

To meet the above requirement for successful operation of a large water distribution system like SSP, it has been planned for flexibility in operation by controlling the volume of water in each pool.

In the controlled volume concept (CVC) of operation, the volume of water between any two cross regulators is altered as desired through proper positioning of flow regulating structures. By controlling the water surface upstream from the control structure at elevations that are higher than required for flow at design capacity, controlled volume operation provides capabilities to pass the design flow through the control structure at any time.

Typically the canal is used for both storage and conveyance. Operation by this method virtually eliminates the need for regulating reservoirs and increases operational flexibility. The response time will thus be limited to a relatively short period required for the transfer of water from the storage pool just upstream of the point of demand instead of from the distant head-works of the canal system. Another advantage is that the pool volumes can be used as a buffer while transforming rapid flow at the downstream end into relatively gradual flow change in the upstream canal pools. Operation by this method will ensure equitable distribution of water to farmers irrespective of their location with respect to head reach or tail reach. A schematic diagram for CVC is given in Figure 2.

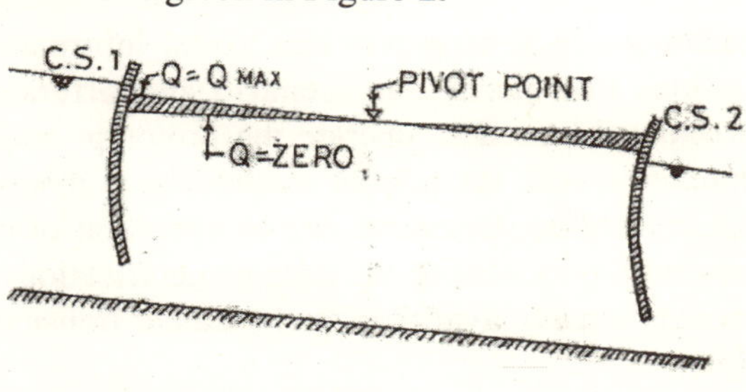

Figure 2: Controlled Volume Concept (CVC)

CVC of operation needs simultaneous operation of control structures and also detailed scheduling of flow requirements at each control structure, which is not possible without Canal Automation.

For the SSP conveyance system, a Constant CVC is to be maintained irrespective of flow variations ranging from design discharge to zero discharge. In "Constant Controlled Volume Concept", when changes in flow are made, the hydraulic gradient line rotates approximately around the water level at the centre of the canal pool. Operation under constant controlled volume concept requires the knowledge of inter-dependence among certain parameters of flow, such as flow in the canal pool, water surface elevation, volume of water in canal pool, gate opening and difference of water levels on either side of control structure.

Implications of CVC for Design of Canal and Canal Structures

In CVC, as the volume in each canal pool is to be maintained equivalent to the design volume at full supply depth, it is possible to design the off-taking canals with higher full supply level. The crest level of head regulators of the off-taking canal could be provided at much higher level than the conventional norms of two third of the full supply depth. Availability of such higher water heads has helped in covering larger areas under gravity flow, thereby realising the objective of extensive irrigation. On the other hand, the peculiar water surface profile rotating around the pivot point has necessitated higher freeboard in the pool length downstream of the pivot.

The Role of Hydraulic Simulation and Mathematical Models

Simulation technique in general provides useful information about system behaviour in a particular scenario and offers a virtual experience without the need to operate the prototype real system. Hydraulic simulation with the help of mathematical modelling is a useful tool for controlling transients due to operation of automatic control devices without violating the design specifications. Various simulation models already available for hydraulic transient studies are listed below:

(1) UCLA	(2) USBR Models	(3) SNUSM
(4) CANAL	(5) CARIMA	(6) MODIS
(7) DUFLOW	(8) PROCAN	(9) DWOPER
(10) NPSSM	(11) NPUSM	(12) MIKE-11, MIKE-21

The different simulation models discussed in detail in "Special Issue of Canal System Modelling", *Journal of Irrigation and Drainage*, ASCE, 1993, No. 119(4), are project-specific. No single model can directly meet the operational requirement of SSP by simulating flow through various types of structures, hence an attempt has been made by SSNNL to develop in-house steady state as well as unsteady state models to meet the specific operation requirement of SSP such as Uncontrolled Canal Siphon (without Gates), Aqueducts, In-line pumping stations, Controlled Volume Concept.

Such specially developed mathematical models provide vital information for application of various principles of control theory to automation of water systems in the absence of a physical model or prototype. With the help of these models, numerical experiments can be done to evaluate various scenarios of canal operation even with different operation and control concepts. It is also possible to know *a priori* which scenario might endanger the structural safety of the canal system. Operators can be trained to develop an insight into canal response by using the mathematical model.

Development of Narmada Project Steady State Model (NPSSM)

The demand of water in a channel varies with time based upon cropping pattern, crop growth stage, rainfall and conjunctive use of ground water in the command. Normally pre-season planning is done and schedules are prepared to cater to the irrigation demands on fortnightly basis, and accordingly flow changes from one fortnight to the next are made in a few hours by operating control structures. However, in-season adjustments in such operation schedules become necessary for a variety of reasons and such mismatches in flow are adjusted once or twice in a 24-hour cycle. After each such change in flow, an unsteady state condition prevails for a specified length of time and then gets stabilised at another steady-state condition. To understand and analyse the behaviour of a canal under steady state, the Narmada Project Steady State Model (NPSSM) has been developed which can simulate the flow in the open channel with various kinds of structure. In this model, depths and discharges at all the nodal points in each canal pool are computed using a gradually varied flow equation. After establishing the initial steady state profile along the canal pool, the water depth in the rectangular section is computed using principle of energy and finally, the gate positions are computed using the gate-discharge equation. Attempts have been made to establish coaxial relation among various hydraulic parameters that govern the flow through a control structure using gradually varied steady state flow conditions.

Development of Narmada Project Unsteady State Model (NPUSM)

Within the realm of irrigation channels, the transient behaviour in an open channel within the canal pools between any two successive control structures is expressed mathematically by the Saint-Venant equations for incompressible flow medium, and the flow through the control structure is given by the gate-discharge equation.

The NPUSM is a one-dimensional numerical model which solves the Saint-Venant equation by weighted four-point implicit finite difference formulation and has been developed as design tool for the canal system of SSP. NPUSM is of a general nature and can be used for specifying boundary conditions either in the form of rate of change of discharge or in the form of gate rate movement. NPUSM is formulated on the basis of CVC and hence the depth range to be maintained at the downstream ends of all the canal pools has to be specified.

Application of NPSSM and NPUSM to SSP Canal

NPSSM and NPUSM are applied to the operation of the first five pools of Phase-I (from 0.000 km to 144.500 km up to the Mahi river crossing) of NMC. Studies with the model have been done considering the Main Canal along with the branch canals lumped at the downstream ends of the canal pools.

Simulation studies have been carried out considering variations in demand to meet crop water requirement, using the actual canal data such as pool length, bed width, bed gradient, Manning's

Roughness Coefficient *n*, side slopes, gate widths, invert levels of the control structures, upstream and downstream reservoir level, gate coefficients, boundary condition for control structure in the form of discharge/gate schedule along with turn-out schedule of flow etc. More than 200 combinations of branch discharges were worked out with the help of the NPSSM model. and coaxial correlation for each canal control structure has been established in the form of a Four Quadrant Graph. The graph depicts relationships amongst differences of head, gate opening, discharge and upstream depth. A typical graphical presentation for Cross Regulator No. 1 is shown in Figure 3. A sample case study output from Steady State Model in tabular form is shown in Figure 4.

COAXIAL CORELATIONSHIP BETWEEN HYDRAULIC PARAMETER
FOR CROSS REGULATOR # 1 @ CH. 13,875 M.

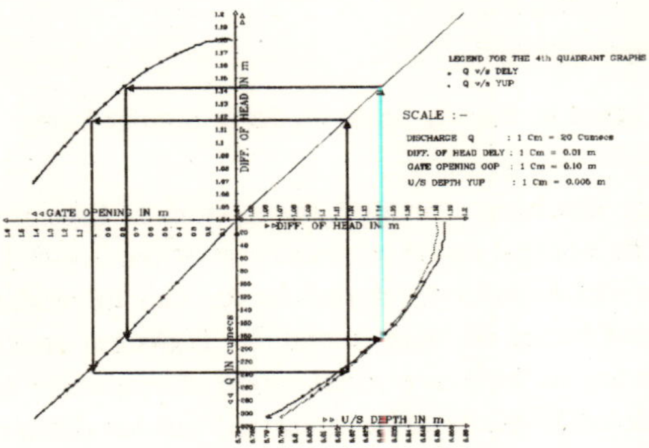

Figure 3: Cross Regulator No. 1

Emergency and Normal Mode of Operation for Narmada Main Canal

In the operation of the canals, two distinct types of control structure operations occur, i.e., emergency and normal mode of operation. Numerical experiments have been carried out using these Mathematical Models for various permutations and combinations of canal discharges. When unexpected situations develop in the distribution system endangering the safety of the system itself, the emergency mode of control structure operation is resorted to whereby

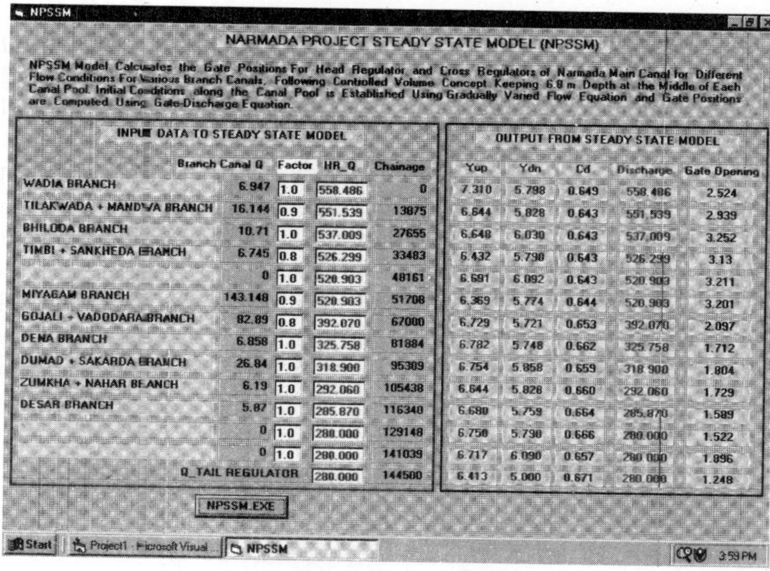

Figure 4: Steady State Model in tabular form

the conveyance discharge through the system is brought to a null (zero) state. The normal mode of operation of the control structures corresponds to situations in which the conveyance through the system is to be changed based on user demands. Transient studies for the case of emergency as well as normal mode of operation for the first five canal pools of NMC have been carried out for different lengths of operation time.

In Emergency Mode of Operation, conveyance discharge is reduced from designed discharge of 1133.0 cumecs to 5.0 cumecs in one hour and the simulation period is six hours. In Normal Mode, conveyance discharge is reduced from designed discharge of 1133.0 cumecs to 50 per cent (567.0 cumecs) of designed discharge in two hours and the simulation period is six hours. The model calculates the time variations of depth and discharge for the given turn-out schedules and for the specified gate movement either in the form of discharge schedule or gate schedule. The model also indicates the exact water surface profile for realistic gate movements. Typical water surface profile upstream and downstream of various control structures and gate movement profile for emergency as well as for normal mode of operation for simulation period of six hour are shown graphically vide Figure 5 and Figure 6 respectively.

Role of Model Studies in Decision Making

- The coaxial relationship among various hydraulic parameters, derived from the Steady State Model can be effectively used as a decision support tool for manual operation as well as for validation of data for automated operation.
- NPUSM permits formulation of efficient rules for normal operation and operation criteria which ensure safe shut-down of the canal system during emergencies.
- It is evident from the graphical presentation of the results, for emergency or normal mode of operation that, increasing the rate of operation of control structure, amount of drawdown on its downstream side as well as the amount of rise in height of surge on its upstream side increases in case of decreasing discharge and *vice-versa* for increasing discharge case.
- In case of stepped operation, either increasing or decreasing the rate of excitation of control structure in steps of its speed, i.e., from 15 cm/min to 75 cm/min for a duration of one minute, the amount of fluctuation of flow variables, i.e. depths or discharges are seen in proportion of excitation, and the following transient effects are noticed.
- Comparison of results of unsteady model with prototype data (California Aqueduct) indicates that results are very close to each other and it can be said that Unsteady Model is quite an accurate model.
- Transient studies carried out with the help of NPUSM model have been quite useful in arriving at the design parameters, i.e., spacing between two cross regulators, amount of lined freeboard and unlined freeboard as well as height of the radial gate etc. have been decided after carrying out simulation studies for emergency and normal mode of operation.

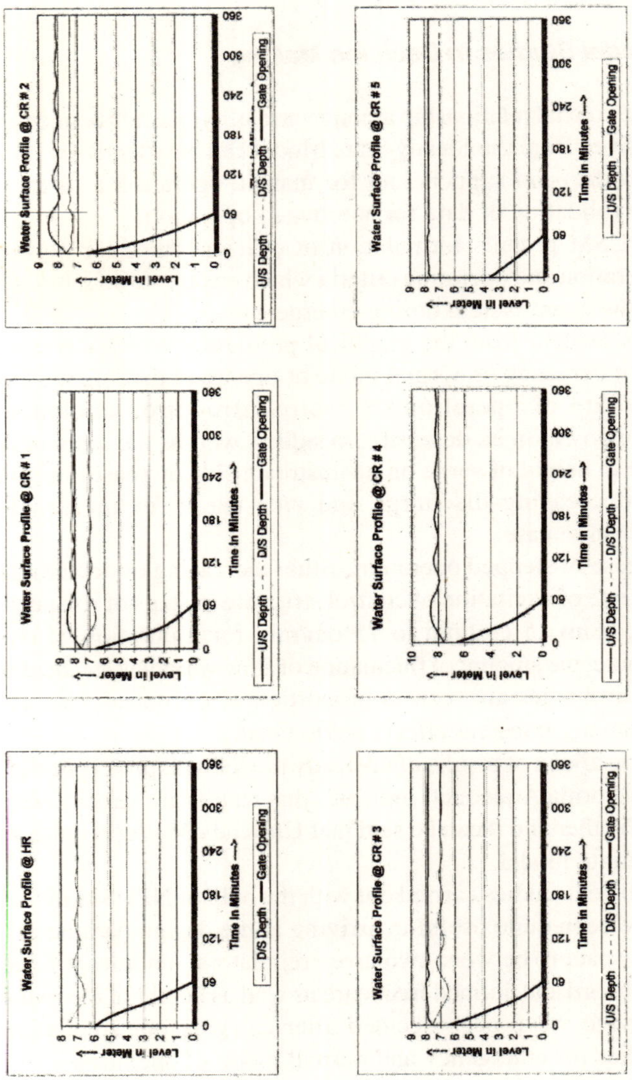

Figure 5: Water Surface Profile and Gate Opening for Emergency Mode of Operation

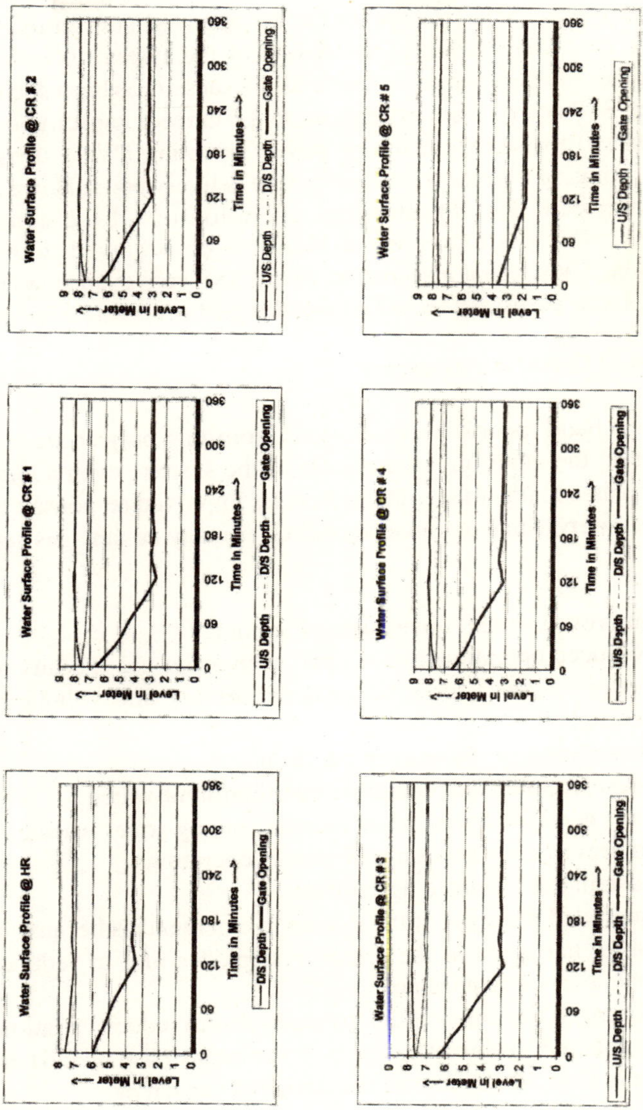

Figure 6: Water Surface Profile and Gate Opening for Normal Mode of Operation

Canal Automation: A Technical Requirement and Not Fantasy

From the foregoing discussion, it is clear that the SSP conveyance system can in no way be expected to be operated conventionally. After careful consideration of the design of the system, functional requirements of operation and other related issues, it has been decided to use the SSP conveyance system (up to 8.5 cumecs capacity in the first instance) for storage also (by way of adopting CVC), thereby virtually bringing the reservoir closer to the delivery point. However, this in turn calls for simultaneous operation of all the control structures, which is extremely difficult to be achieved without automation. Thus, canal automation in the SSP conveyance system is evolved as a necessary technical solution and is no way a luxury.

Benefits of Canal Automation

On the one hand, people's expectations from this project are very high, and on the other, given the scale of the system, even a slight mismatch can endanger the system. Besides improved response time and better safety, canal automation will offer many advantages such as:

- Improved water conveyance efficiency.
- Reduced operational losses and even water use (because of better response time, better dependability and volumetric billing).
- Realisation of 'more crop per drop'.
- More accurate and equitable distribution of water.
- Less need for operation of escapes, saving water wastage.
- Improved coordination with power operations.
- Sustainable irrigated agriculture.
- Improved response of the system to emergencies because of better harmony in the entire system, leading to reduced maintenance requirements.
- Enhanced flexibility of the system to absorb sudden variations in demand. In many cases such variations (rise or fall) are more than 50 per cent of the designed capacity.
- Continuous log of the events would be possible, which can help build the knowledge base for future.

Evolution of Operational Plan

The present plan of operation for the SSP canal conveyance network is an outcome of a continuous evolution process going on since the early 1980s. The World Bank Experts during their initial discussions (1980-1983), expressed the view that the conveyance network design should be based on CVC of operation which would minimise response time in adjusting to anticipated rapid changes in network flows for safe and efficient operation. WB also suggested, and it was finally decided, that conveyance network designs would be analysed and checked by state-of-the-art Hydraulic Transient (wave movement) methodologies, including computer based models for analysis of normal and emergency canal operations, analysis of wind run up, and other hydraulic and physical consideration. Further, the WB Mission in May 1981 desired urgent investigation of operational control of canal network to ensure high overall conveyance efficiencies.

The document of the World Bank, Staff Appraisal Report, India—*Narmada River Development, Gujarat Water Delivery and Drainage Project, February 12, 1985*, laid down the Standards and Criteria for Planning, Design, Construction and Operation and Maintenance of the Irrigation and Drainage and M and I water Supply facilities of SSP.

It was thus decided to base the design for SSP conveyance network on CVC of operation and also to analyse and check it using state-of-the-art Hydraulic Transient (wave movement) methodologies, including computer based models for analysis of normal and emergency canal operations. Professor J.S.R. Murthy, Civil Engineering Department, IIT Bombay, was appointed as a Part-time Consultant for the development of the Canal Simulation Model. Professor H.B. Dave, L.D. College of Engineering, Ahmedabad was appointed as a Part-time Consultant for Planning and Design of Operational Controls and Transient Studies for Narmada Canal System. Various training programmes on Semi-Automated Canal Operation as well on Canal Automation were organised by design organisation for imparting exposure to the participants of SSNNL.

The Preliminary Feasibility Report on Control and Communication System of Narmada Canal Network was prepared by GCEL and discussed. It was suggested that assumed values of data scanning rate for communication and sensor resolution be reviewed using the NPUSM simulation programme. Further, following the simulation studies carried out with NPUSM with single pool data for different modes of emergencies, scan rate proposed was two minutes. GCEL submitted Final Feasibility Report on *Control and Communication Network for Narmada Canal Project* in April 1990.

Various simulation models for canal operation were discussed at length and owing to several advantages of the in-house model, NPUSM was recommended to be used after comparison of data with prototype results. During the World Bank Review Mission of May 7-14, 1990 at Vadodara, the NPUSM was discussed with Mr. D.C. Rogers, Expert on Canal Transients. His observations are listed below (Ref. Sardar Sarovar Water Delivery and Drainage Project (Cr. 1553-IN) From Michael Baxter To Chairman, SSNNL Dt. July 3, 1990).

> "The canal unsteady flow model NPUSM has been developed by Dr. Murthy and Associates at the Nigam, although its general capabilities are yet quite inferior to many other models in U.S. and Europe, NPUSM appears to be Technically Sound and has apparently been adequately Validated with Comparison to other Models and field data. NPUSM has the advantage of customisation to SSP needs, which will be valuable as future development is required."

In November 1992, SSNNL invited bids for providing consultancy services for the Remote Monitoring and Control System for SSP Water Conveyance System. In the process three project engineers—the two authors of this chapter and V.M. Rana–were deputed for a Ph.D. Programme under external registration scheme at IIT Bombay under the guidance of Prof. J.S.R. Murthy, IIT, Bombay and Mr. M.M. Shah, Executive Director, SSNNL. Mr. Acharya for his thesis entitled *Development of Downstream Control Algorithm* and Mr. Joshi for his thesis entitled *MIS and Simulation Based Approach for Irrigation Operation and Management* were awarded Ph.D. degrees in 1994-95. Mr. Rana was awarded a Doctorate for the Development of Auto-Controller, in 2001. At various points of time, project engineers also visited projects such as California Aqueduct, Central Arizona Project, Delta Mendota Canal (USA), Canal de Provence and Rhône River Scheme (France), West Nubaria Project (Egypt) and Canal de Rocade (Morocco).

Operation Plan for SSP Conveyance System

For the purpose of irrigation management the gross command area of 3.428 million ha of SSP is divided into blocks of approximately 26,000 ha. The irrigation management data for each block will be collected by a Data Collection Centre (DCC). The demand of water will be collected from the co-operative societies of the farmers known as Village Service Area (VSA) at least 24 hours in advance. For collection of these demands and compilation, DCCs will be provided in the command area, which will also serve as nodal centres for all the dealings with the VSAs including billing of water charges. It is planned to have 132 DCCs. These will communicate data required for regulation of branch canals to Divisional Operation Centres (DOCs), which will be in charge of branch canals. Considering the limitations from hydraulic and canal operation points of view, 15 DOCs will cover the command area. They will report to the Main Control Centre (MCC) located near Gandhinagar, which will be in charge of operation of the Main Canal.

There are about 1,400 regulating structures on the canal network with up to 8.5 cumecs discharge carrying capacity each. A "Distributed-Decentralised" approach has been adopted for monitoring and control. The whole automated canal network will be monitored and controlled by a three-tier hierarchical network of control centres—MCC, DOC and Remote Terminal Unit (RTU). The RTU will receive the data such as Water Level and Gate Position from various sensors, communicate it to the DOC, receive the command signals from DOC and pass them on to the control cabinet to actuate the gate hoist motor for raising or lowering the gate. The DOC will derive the gate schedule corresponding to the discharge requirements received from DCCs and the allocation policy. The MCC will receive gate schedules and corresponding discharge requirements from all DOCs for various branches. Taking into consideration the availability of water in the reservoir, priorities, policies and other related factors. MCC will approve water allocation to various DOCs and simulate the CR/HR gate schedules for the main canal. The approved branch canal gate schedule along with the corresponding main canal HR schedule will be conveyed to the respective DOC. A schematic diagram representing Remote Monitoring and Control System for a typical structure is shown in Figure 7.

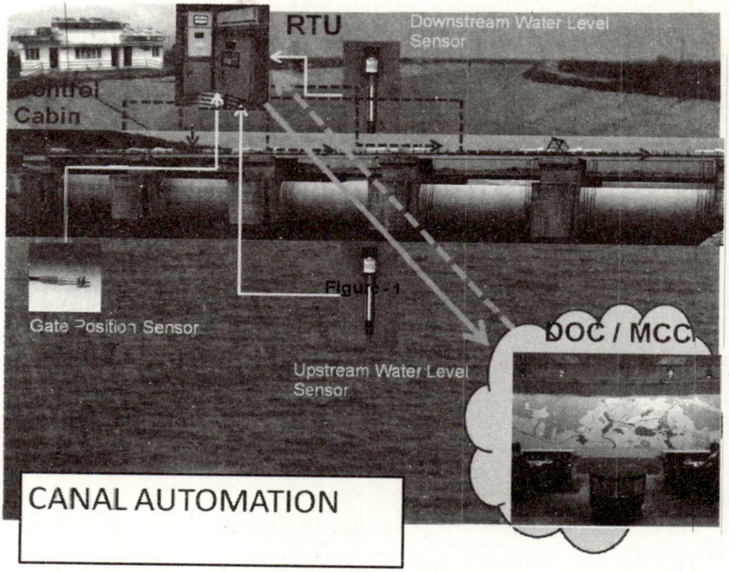

Figure 7: Remote Monitoring and Control System for a typical structure

Communication System

Communication (information acquisition and dissemination) is important for managing any canal network. In SSP with automated canal operation it becomes crucial. A reliable transmission channel is a must for automated canal operation and inter/intra-group field communication. The requirements of the communication system are as follows:

- Data communication between RTUs and operation centres for control and monitoring.

- Data and Voice communication between villages (VSAs) and the DCC for irrigation management operations.
- Voice communication between sectional officers and operators of manually controlled structures.
- Voice and data communication among officers at various levels for O&M of canal system.
- Video communication between RTUs and DOC as well as DOC and MCC

For the automation net, optical fibre cable will be used with VSAT based back-up communication at the critical links. For regional net (other than the data communication among control centres) TDMA terrestrial net/Radio Trunking/VSAT for voice and data communication is planned.

Canal Automation: An Interdisciplinary Application

Remote Monitoring and Control System is a multi-disciplinary technology which needs technical expertise in various disciplines like Civil, Mechanical, Hydraulic, Electronic, Electrical, Software and Hardware, Instrumentation and Control and Communication Engineering, and in Computer Science.

Canal Automation has been successfully implemented in the following projects in various parts of the world:

Country	**Nae of the Project**
Egypt	West Nubaria Project
Morocco	Canal de Rocade
France	Canal de Provence
	Rhône River Scheme
	Durance-Verdon Multipurpose River Development Scheme (EDF)
Netherlands	Flevopolder Project
USA	California State Water Project
	California Aqueduct
	Delta Mendota Canal
	Central Arizona Project

In India too, attempts have been made and are being made to automate canal operations in the following Projects:

- Chambal Project (Madhya Pradesh)
- Khadakwasla Project (Maharashtra)
- Majalgaon Project (Maharashtra)
- RAJAD Project (Rajasthan)
- Tungabhadra Project (Karnataka)
- Indira Gandhi Nahar Pariyojana (Rajasthan)
- Parambikulam Aliyar Project (Tamilnadu)

Canal Automation Pilot Project: Current Status

Canal Automation has been planned as an integral part of the Planning and Designing of the SSP Canal network and it has been decided that any part of the network with discharge carrying capacity above 8.5 cumecs will be covered under Remote Monitoring and Control System. As canal automation on such a large scale is being implemented for the first time in the world on the SSP Conveyance System, to obtain first-hand experience it has been decided to implement it as a Pilot Project on about 600 km length in Phase-I of the command area, consisting of NMC Reach km 0.0 to 458, Vadodara Branch Canal (with Jambusar and Kundhela Branch Canals) and Sakarda Branch Canal systems.

Implementation of the Canal Automation Pilot Project was deferred in November 2005, for want of full development of command area. CAD works as well as Gate works of Branch canal, distributories, minors and sub-minors for the pilot project are planned to be completed by December 2011. Therefore, it is now decided to go for implementation of Pilot Project for the Canal Automation and Telecommunication Consultants India Limited (TCIL), New Delhi has been awarded the Consultancy Assignment for Reviewing Detailed Design and Bid Documents for Canal Automation Pilot Project, prepared by SCP-GERSAR in year 2005, and preparing the final version Bid Documents so that implementation of the pilot project can be taken up. On receipt of the final version of Bid Document from TCIL, Tenders for implementation for Canal Automation Pilot Project will be invited in the first quarter of 2011.

Thus, during 2011-2013, the Pilot Project of Canal Automation is planned to be implemented in phases. After gaining this experience it will be scaled up to the entire network above 8.5 cumecs with suitable adaptations and modifications as may be necessary.

Beyond the Horizon

Automation of operation of the SSP canal conveyance system is not an end in itself, but will be a means to begin a much more important phase of the project. Installation of the state of the art instrumentation, control equipments communication system, application software etc. and their integration would not only provide unique opportunities for project engineers to develop their skills, but would also serve as the largest prototype laboratory of the world. Field experiments with unlimited permutations and combinations would be possible with an extensive documentation facility. Results of such studies could be far more important for the engineering fraternity to better understand the canal operation. The contribution of SSP on this count will be no less than that of a technological university, and when this is realised, it would be far beyond the original dream of the experts who had sown this seed.

Last but not the least, automation of the SSP canal conveyance system needs to be viewed as the means and perhaps the only means to achieve the mission of this project which is to harness the water of the mighty Narmada for the survival and development of millions in a sustainable way. After successfully addressing the need of the Remote Monitoring and Control System on the SSP canal conveyance system with discharge capacity above 8.5 cumecs, linkages with the distribution network and water users' associations would be possible. An environment providing opportunities for seamless integration of all the stakeholders through soil health cards, smart cards for the farmers (*Kisan Cards*), automated farm level operations based on real-time moisture conditions in the crop root zone, marketing cooperatives, banks, crop yield insurance etc. can no longer be ruled out as a fiction. To aim thus is to see the benefits of the technology reaching out to a common farmer in the remote rural areas, where the Narmada flows.

31

Connecting the Last Mile

S. Jagadeesan

SSP—The Saviour amidst Sobering Scarcity of Water

By 2030 the global demand for water will be 40 per cent higher than it is today and more than 50 per cent higher in the most rapidly developing countries. The drivers of this resource challenge are fundamentally tied to economic growth and development. Out of the total global water withdrawal of 4,500 billion cubic meters (BCM), agriculture accounts for approximately 3,100 BCM (71%). Even in the year 2030, it is going to remain dominant with 65 per cent, when the overall demand would have touched 6,900 BCM. At that time, India's demand for water will be about 1,500 BCM, of which 1,195 BCM (80%) is estimated to be agricultural, driven by domestic demand for rice, wheat and sugar for a growing population (The 2030 Water Resources Group, 2009). The projected water scenario of Gujarat for the year 2025 is more skew, with 46.258 BCM agricultural water demand (out of a total demand of 50.43 BCM)—that is 91.72 per cent (Tahal, 1987; BIG 2020). The dwindling surface water resources of Gujarat state with intra-state imbalances might have prompted the then Director General of International Water Management Institute (IWMI) in September 2001 to cite the case of Gujarat as a 'quite typical example' in his keynote address at the International Conference on "Sustainable Food security for All by 2020" (IFPRI, 2001). But there are many other reasons that make water management in Gujarat more challenging—be it urbanisation, industrialisation or any others with a thirst for water. And it is in this backdrop that the Sardar Sarovar Project (SSP) has emerged as saviour, even at its partly implemented stage.

From Conceptualisation to Implementation—A Long Journey

While Gujarat celebrates its golden jubilee, in April 2011, SSP will also complete 50 years of its long journey. The foundation stone of this mega-project was laid on April 5, 1961, but the actual construction could commence only in 1985 after the resolution of inter-state issues through a constitutional framework. The conclusion of the high-level committee headed by A.N. Khosla (September 1964) did not put an end to the inter-state dispute for sharing of Narmada water and a series of meetings between representatives of the state governments failed to break the deadlock. Therefore, in July 1968, Gujarat sought the appointment of a tribunal under the provisions of the Inter-State River Water Disputes Act (1956), to adjudicate the dispute. Finally, the matter was referred to a special panel constituted for the purpose—Narmada Water Dispute Tribunal (NWDT) in 1969—around eight years after the foundation stone was laid. The time the tribunal took to give its final verdict (December 1979) was even more than that—about 10 years.

The NWDT award cleared the decks for advancement of the project, but it was just the beginning of a process to get necessary approvals from different authorities. Before actual commencement of the work, the Planning Commission was required to check the economic and financial feasibility of the project and approve the cost estimates. Similarly, the relevant Union ministries were required to assess the environmental impact of the project and allow utilisation of forest land for non-forest use. While the Planning Commission gave its approval to the cost estimate in October 1988, a conditional environmental clearance was given to the project in June 1987.

Even after commencement of the dam construction work, a series of lengthy delays continued to affect its progress in a most destructive manner. Instead of affecting the work in stages, if all the delaying factors had shown their impact at one go, the project might have been completed in the 20th century itself. With the progress of work, changing stances became imperative for the project opponents. First they were agitating only for the proper resettlement and rehabilitation

(R&R) of the displaced families. But after commencement of the work, they started demanding shelving of the project and even raising doubts over its economic viability and environmental and archaeological aspects. This was clearly indicative of their real motive. The tactics worked at least in the short-term and under mounting pressure. Following the Madhya Pradesh (MP) State Assembly's direction, the Narmada Control Authority (NCA) took a decision in January 1995 not to allow construction of the dam above 80.30 metres height. Acting on a writ petition filed by the Narmada Bachao Andolan (NBA) in April 1994, the Supreme Court stayed further construction of the dam in May 1995. Even dam-related works other than the spillway were not allowed through an order in April 1996. The stay was withdrawn only in February 1999 through an interim order allowing raising of dam height up to 88 metres. In its verdict given on October 18, 2000, the apex court allowed dam construction up to the desired height subject to compliance with environmental issues as required under the Ministry of Environment and Forests (MoEF) clearance conditions and the provisions for rehabilitation by the NWDT award.

Subsequently, a concerted strategy and satisfactory compliance with obligations in terms of rehabilitation of project affected persons (PAPs) and environmental protection measures, ensured that the dam height could be raised up to 100 metres (2003), 110.64 metres (2004) and 121.92 metres (2006). The last mentioned height has made available live storage of 1565.8 million cubic metres (MCM—1.27 million acre feet)—more than the combined live storage of Panam, Damanganga and Shetrunji reservoirs in the state and even more than the live storage capacity of Krishnarajasagara (Karnataka) or Idukki (Kerala).

Some of the other factors that delayed project execution were the lengthy process of land acquisition, clearances from various authorities for utility crossings on canals, demand for more communicating and cross-drainage structures on canals by local farmers, abandonment of work by several agencies and change in alignment of canals and changing soil profiles. Construction of Minors and Sub-minors in particular becomes more challenging and time consuming because of smaller quantities of work to be executed in geographically spread-out areas in which more stakeholders are involved, and even access to the sites may be difficult for construction machinery. On the other hand, it becomes vitally important to realise the irrigation benefits by supplying water to farmlands. In this perspective, the remaining works of SSP can be viewed as the last-mile connectivity, essential to meet the overall objectives.

Present Status of the Project

With a very modest beginning diverting just 705 MCM water in the canal system in 2001, the diversion spectacularly increased to 6194 MCM in 2004 and during 2009 it was 5870 MCM. By the end of December 2010, the electricity generated by two hydropower stations with installed capacity of 1450 MW was 17,924 million kWh.

Overcoming all hurdles, to quench the thirst of Gujarat and Rajasthan, the 458 km long Narmada Main Canal (NMC) with its 630 structures has been completed and water has been flowing all through since March 2008. It is proposed to use this canal not only for conveyance but also for storage of water to improve the response time of the system. The NMC alone has the capacity to store 220 MCM of water.

A canal network of length 19,248 km has already been completed by the end of December 2010. This gives an impression that only one fourth of the canal network has been completed, but even this achievement is not small by any means. It is more than eleven times the distribution network (up to minors) envisaged in Indira Sagar Project. With its already constructed 26 branch canals and several distributaries, the Narmada canal has become a perennial water grid for Gujarat, which otherwise has only eight perennial rivers (out of a total 185), and all except the Mahi concentrated in south Gujarat, leaving three-fourth of the state dependent on artificial water distribution systems like Narmada canal. Through this canal system, water from south Gujarat transferred to parts of north Gujarat, Saurashtra and Kachchh has already started reducing the water supply-demand imbalance due to uneven endowment of water resources across the four regions of the state. With nearly 300 MCM supplied to Rajasthan through the NMC, farmers in one of the driest regions of that state irrigate nearly 60,000 ha of land through sprinkler systems using Narmada waters.

Considering the canal network upto minors, irrigation potential developed is 5.03 lakh hectare—almost six times the ultimate irrigation potential of Dharoi Dam, four times that of Indira Sagar Project and three times that of Ukai! Phase-wise status of the command area development is shown in Figure 1.

SSP is all set to contribute—after completion of the ongoing Major and Medium projects—1.3 million hectares to the national goal of creating irrigation potential of 4.2 million hectares of land under the ambitious Bharat Nirman Yojana. This would probably be the largest single contribution in the country.

Benefits of Narmada water have already reached 6,513 villages and 114 urban centres in the state. As a result the number of villages being supplied drinking water through road tankers has reduced significantly.

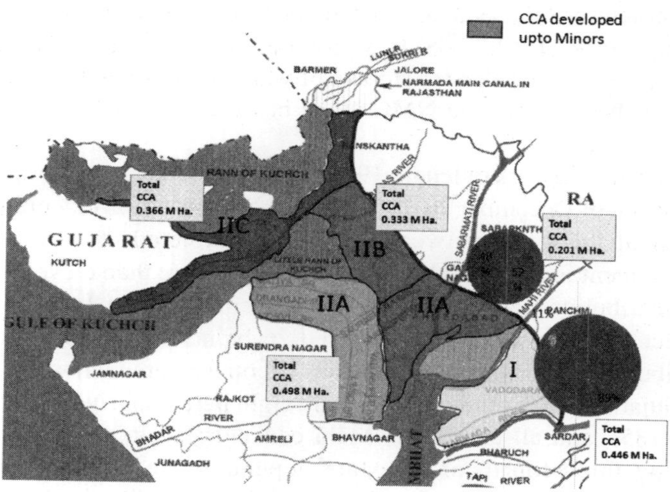

Figure 1: Phase-wise Status of Command Area Developed up to Minors

SSP is the first project for which Environment Impact Assessment (EIA) was carried out prior to approval of the project. The challenging task of analysing the several recommendations made in the EIA studies and formulating a comprehensive plan was attended at the beginning. The planners had identified the areas of impact—Forestry, Agriculture,

Wildlife, Health and Fisheries. SSP has already completed treatment in 29,157 ha of the catchment are lying within Gujarat. Compensatory afforestation in 13.950 ha of forest and non-forest area and Dam vicinity plantation in 550 ha have been completed. More than 4,300 ha have been afforested so far in the completed reaches of the canals.

Ensuring *pari-passu* R&R and implementation of ameliorative and compensatory measures for environmental protection with the increase of the height of the dam, the construction of the dam has been taken up as per the Award of the Tribunal, subject to clearance of the R&R Sub-group and the Environment Sub-group of the NCA. Each State is required to abide by the terms of the action plan so prepared by the NCA and in the event of any dispute or difficulty arising, representation may be made to the Review Committee of NCA (RCNCA). State-level Grievance Redressal Authorities (GRA) have been formed to ensure effective implementation of the policy.

Socio-economic Impacts

Even with partial completion of the dam and canal network, Narmada water has been playing its role. Its value as a harbinger of socio-economic and human development is reflected in its various benefits, tangible as well as intangible, ranging from rural water security to combating global warming. Narmada water has significantly contributed to the consistent performance of Gujarat in the agriculture sector, helping the state recover from a major shock due to a steep fall in farm outputs following the severe droughts of 1999 and 2000 (Kumar *et al.*, 2010). With the partly constructed canal network, inter-basin transfer of Narmada water to twelve major rivers has been realised and its impacts are recorded in terms of augmented recharge of groundwater in the alluvial areas of central and north Gujarat, improved quality of groundwater for drinking and its associated health benefits, primary education and so on.

There are unarguably some delayed benefits, but the benefits already available are also noteworthy. Cumulative eco-friendly generation of more than 18,000 million units of electricity has put SSP prominently on the world atlas of hydro-power projects. In the five years from 1996-97 to 2000-01 power consumption for agriculture in Gujarat had risen by more than 50 per cent from 10,105 MUs to

15,489 MUs. Its percentage share in the total consumption of electricity in the state also rose from 35.98 per cent to 45.12 per cent. Miraculously, after the operationalisation of NMC the electricity consumption for agriculture came down from 45.12 per cent in 2001 to 21.10 per cent in 2009. In absolute terms the consumption went down from 15489 MUs to 11733 MUs. This miracle was due to two interrelated reasons: (1) the release of water into the dry river beds of north Gujarat with deep sand deposits raised the groundwater table through heavy percolation; and (2) the pumping of groundwater from deep aquifers reduced with the introduction of replacement water from canals and rivers. It is worth noting here that Gujarat's annual subsidy towards electricity consumed by the agriculture sector is of the order of Rs. 30 billion. Irrigation with Narmada water has brought relief from this due to reduced dependence of groundwater reduced pumping costs in the areas where groundwater recharge has taken place. Ex-President A.P.J. Abdul Kalam, when launching the book *Convenient Action: Gujarat's Response to the Challenges of Climate Change*, mentioned this and said "... Besides generation of power, it has also brought about a saving of over 50,000 million units of electricity which would have been otherwise consumed to draw water from ground sources. This has brought down the electricity consumption for agriculture from 45 per cent to 21 per cent in the period from 2000 to 2009. This energy generator and energy saver is an example for the entire nation." And this shows what role the SSP has played.

National Remote Sensing Agency (NRSA) has estimated the area irrigated by the Narmada canal network at 392,158 ha in Phase-I and 267,430 ha in Phase-II. According to this study, more than 660,000 ha land is being irrigated regardless of the missing links in the command area. Noted Economist and former Union Minister Y.K. Alagh, who was also instrumental in planning the project, has expressed this achievement in these words "... With more irrigation, total yield goes up. As ISRO satellites show us on January 31, 2009 area irrigated by SSP canals was 3,74,606 hectares. On September 28, 2008 it was 2,46,995 hectares. SSP is delivering to farmers, with good and bad canals, 6.21 lakh hectares of irrigation. Of course we want thrice as much and we want it controlled, so that more people benefit, our land is not spoiled and our crops are diversified. But,

Planning for Prosperity said in 1985 Gujarat's agriculture will grow by 6 per cent after it delivers. That's the way it is today" (Alagh, 2010).

Results in the area under coverage indicate that Narmada water has not only provided irrigation facility in the said area, but also facilitated crop diversification and enhanced yield. The May 2009 Report of International Water Management Institute (IWMI) and International Food Policy Research Institute (IFPRI), titled *Agriculture Performance in Gujarat since 2000: Can it be a Divadandi (Lighthouse) for other States?* took special note of Gujarat's 9.6 per cent growth rate of GSDP from Agriculture, which is the highest amongst Indian states. This eye-catching performance may be attributable to a whole array of factors like increase in area under irrigation, crop diversification, i.e. more commercial crops like cotton, vegetables, cumin etc., increase in productivity, i.e. yield due to HYV seeds, improved fertilisers, pesticides, better on-farm management of resources, better agricultural financing, improved availability and coordination of labour or a combination thereof. Nevertheless it is acknowledged by many that Narmada water *inter alia* has made all other efforts more meaningful. The conclusion of a study conducted by Sneh Setu Foundation, Vadodara on 'Impact of Sardar Sarovar Project Water on Selected Villages of Gujarat' proves to be some startling revelations. The study report, released in September 2008, says that yields of various crops have increased two-to five-fold after Narmada water was made available for irrigation. Similarly agricultural income per *bigha* has also risen two-to four-fold. The farmers in the region are able to sow winter crops and switch over to high-value crops after assured water supply became available through the canals. It adds that work opportunities for farm labourers are on the rise and their wages have increased by 40 per cent to 50 per cent amid acute shortage of such workers, obviously due to availability of Narmada water for irrigated crop production in winter. The same study notes that introduction of Narmada water had huge impacts on access equity in groundwater. The monopoly power of tubewell-owning water sellers in the area has drastically reduced, as reflected in the steep fall in charges levied by them for irrigation services.

The respite brought by Narmada water is reflected in the number of incremental tubewell installations decreasing from 1,146 in 2000-

01 to 47 in 2008-09. Similarly number of villages getting water supply by road tankers got reduced from 4,054 to 326 during the same period.

Lamenting at the attitude of the project's opponents, well known journalist B.G. Verghese had rightly stated:

> They have revelled in scoring small points but have missed the wood for the trees. Whenever benefits were within reach, with every raising of the dam, they have tried desperately to halt progress. The theory they cling to is that the Sardar Sarovar Project does not benefit anyone, so the realisation of benefits must be stopped at any cost. Once benefits flow, the NBA and its friends will be silenced and out of business. Indeed, drinking water is already being supplied to 25 million people. There have been significant improvements in health, women's time disposition, distress migration, livestock, agricultural productivity, incomes, employment and land values and living conditions (*Outlook*, May 3, 2010).

Considering the tangible and intangible benefits of Narmada water such as stabilising and boosting agricultural production, creating employment in rural areas, prevention of migration and rapid desertification, arresting groundwater depletion and improving groundwater quality, reducing infant mortality ratio and school drop-out ratio and improving overall public health, it becomes the responsibility of the Project authorities to expeditiously complete the remaining works so that such benefits can reach to the remaining parts as well.

Strategy for Connecting the Last Mile

SSP has successfully adopted path-breaking approaches for arriving at state-of-the-art technological solutions at every step of its implementation. It has been a forerunner in terms of offering new concepts and evolving fresh engineering standards—be it planning, designing or implementation. Engineering issues dealt with in the different project components like dam, hydropower, canal, canal structures and so on, had no off-the-shelf solutions existing anywhere in the world. "Necessity is the mother of invention"—this has been absolutely true in the case of SSP and has led to many distinctive elucidations. Some of the unique features built into the planning of the project are concrete-lined canals to reduce conveyance losses, use of control volume concept for design of distribution systems, efficient water use allocation with optimised crop planning for 13 different agro-climatic zones of the command, extensive irrigation to a 180,000 ha area with just 21 "delta supported by the use of 8" of groundwater, deepening of village tanks for borrowing soils for canal embankments, computerised automated operation of canal system, participatory irrigation management through Water Users Associations, and promoting micro-irrigation systems like drip and sprinkler for efficient water use. In order to complete the project and the command area within the time schedule, the following measures have been adopted leading to good governance.

The Project Authorities have collectively taken certain strategic and policy level initiatives towards "Target on Time" so as to complete the project as planned and to deliver water to the remaining parts of the command area at the earliest possible and in and the most effective manner. The Authorities are of the view that with all such initiatives and with the active participation of all the stakeholders, the project objectives would be realised sooner by speedy execution of various actions.

The Road Map

An ambitious and yet achievable road map has been prepared with an overall target of completing the Project by 2014-15. This includes the remaining work of the Main Dam and works of Branch Canals, Distributor Canals, Minors, Sub-minors and the structures thereon. This in turn calls for a massive construction programme in the next three years, for which adequate financial resources, contractual framework, land acquisition, designs, project monitoring, quality assurance and control have been planned in advance. A policy decision has been taken to implement the canal network in such a way that it leads to successive development of the command area and not as scattered works in different canal segments. Wherever possible, the works of delivery distribution network are also being initiated concurrently to ensure that water reaches the farms through the planned canal network.

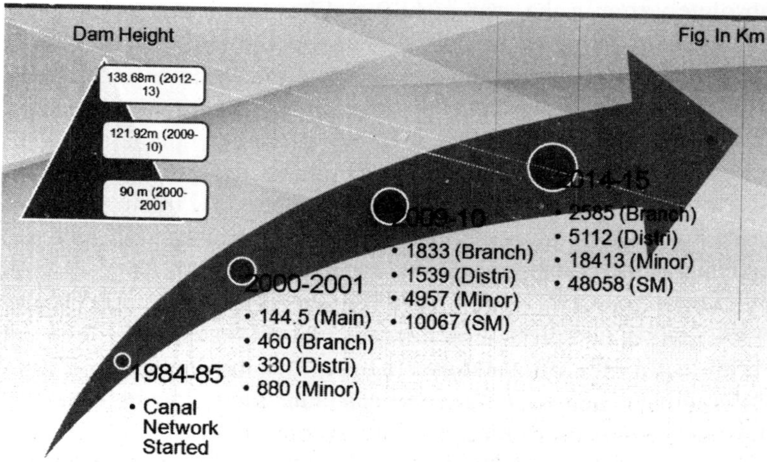

Command Area Development

Importance of command area development activities cannot be overemphasised because of their role in realising real irrigation benefits. The SSP command area fully developed to Sub-minors level is around 348,000 ha. Over 1.5 million ha command area is yet to be developed to this level, which is challenging by any standards despite the fact that another 155,000 ha area has been developed to Minors level. Year-wise and area-wise detailed planning has been done to complete the remaining CAD works and their progress is being monitored. Experience has shown that due to missing links the downstream command remains deprived of water. Such missing links may be due to non-availability of land, road, railway or other utility crossings, crossing of any existing canal of other medium/minor project etc. Efforts are therefore on to identify and remove such missing links and 'actual water flow' is made the target rather than misleading percentage progress in terms of physical quantities. A comprehensive implementation plan encompassing various CAD activities is being prepared for each area. CAD activities generally considered include construction of water courses (Sub-minors) and field channels; land levelling and land shaping; provision of field drains; assisting the farmer in development of ground water resources and conjunctive use of surface and ground waters; provision of a network of basic infrastructure roads and farm roads; developing, marketing and warehousing facilities; establishing agricultural research-*cum*-demonstration farms; strengthening of cooperative structure for input and credit facilities to farmers; arrangements for timely supply of inputs like seeds, fertilisers, pesticides and consolidation of landholdings where necessary.

Expediting Land Acquisition

The land acquisition for canals is to be carried out under the Land Acquisition Act. The land required so far was in the South and Central Gujarat, where prices are higher than those in the command areas of North Gujarat, Saurashtra and Kachchh. Most of the land acquisition now required is in the areas of North Gujarat (Banaskantha and Patan districts), Saurashtra and Kachchh region, which are part of the drought prone areas, and where land holdings are generally larger. With low to medium rainfall, high annual variation and semi-arid to arid climate (IRMA/UNICEF, 2001), enhancement of agricultural production in this region depends heavily on the availability of water. The farmers are volunteering to hand over land for construction of canals so that they can avail of irrigation benefits at the earliest. With the cost of groundwater irrigation becoming astronomical and agricultural land getting degraded owing to the use of high TDS groundwater for irrigation, these farmers are the victims of environmental degradation caused by excessive mining of groundwater, and Narmada water could address these problems to a large extent. Besides this, it has been decided to pay the land acquisition compensation at the *Jantri* Rate (official market rate) and acquire land through mutual consent, so that it may become available earlier and the farmers get fair compensation without resorting to court intervention. A High Level Committee headed by the Revenue Minister is steering and guiding the land acquisition process. The committee meets every week and several meetings are also held at different district and taluka headquarters. This initiative has helped address the land acquisition related challenge adequately and in with participation of the beneficiaries. From the seven districts of North Gujarat and Saurashtra and from Kachchh district, around 10,000 ha of private land has been acquired which has facilitated execution of canal network construction in these regions.

Outsourcing of Joint Measurement and Preparation of LAQ Proposals

As per the land acquisition procedure, a joint measurement survey needs to be undertaken in six times the land area actually required to be acquired. With a view to expedite this, a conscious decision was taken to outsource this activity instead of carrying it out departmentally. This practice is being followed since over a year and it has proven to be very successful in expediting joint measurements of the land proposed to be acquired. Around 60 professionals are working on this aspect using the latest total station technology. As a result joint measurement survey in 221,950 ha area is complete and only 32,000 hectare area remains to be measured. This has helped in the preparation of Land Acquisition proposals and further processing.

Outsourcing of Survey and Design Works

For design of the Main Canal, Branch Canals and Sub-Branch Canals various types of control, communication and cross-drainage structures, a centralised in-house design approach has been adopted and more than 500 structures have been thus designed during last year. About 95 per cent of the designs are complete for all the Branch Canals except Kachchh Branch Canal for which the design work has been partly outsourced to RITES Limited. For the distribution network, a major constraint faced during implementation was the inability to provide designs to the contractors in time, resulting in delay in implementation. SSNNL, therefore, decided to outsource this task. For expeditious completion of Network Planning including designs of Distributaries, Minors and Sub-Minors, in-house efforts are being supplemented by consulting organisations like WAPCOS. For major works like pumping stations on the Saurashtra Branch Canal and Kachchh Branch Canal, organisations like NTPC and Tata Consulting Engineers have also been consulted.

Initiating New Works under Item Rate/EPC Contract System

For completion of the remaining works of Sub-branch canals of Saurashtra Branch, namely lining of the Morbi, Vallabhipur, Dhrangadhra and Limbdi Sub-branches, and distribution network in

Maliya, Vallabhipur, Dhrangadhra, Limbdi, Botad and Morbi branch canals, 28 tenders worth around Rs. 15 billion have been invited, which are being finalised after evaluation. This will help in irrigating over 500,000 ha command area of the drought prone Saurashtra region.

For want of detailed geotechnical investigations, designs and network planning some of the works of Kachchh branch canal were originally planned to be taken up after two to three years. However, recognising that many experienced project engineers are due to retire, it was felt necessary to synchronise the works of the Kachchh region with those of Saurashtra. For bringing it forward and to further expedite the work progress in developing Command Area, the EPC (Engineering, Procurement and Construction), EPC mode of contracting is now resorted to, which will include survey, geotechnical investigations, design of canals and canal structures and network planning. In the case of Kachchh Branch Canal, works on the initial 45 km length are in progress including distributaries off-taking from 0 to 30 km of KBC. For the remaining length of KBC up to 360 km, six EPC packages are designed, of which four, worth around Rs. 13 billion are in the final stage of award. The following Table shows the magnitude of these packages in physical and financial terms.

Name of the Package	*Estimated Cost in (Rs. crores)*	*Earthwork Quantity (Lakh m^3)*	*Concrete Quantity (Thousand m^3)*
KBC Ch. 45.00 km to 54.90 km	277.97	39.17	250.0
KBC Ch. 65.00 km to 82.30 km	223.49	89.55	257.2
KBC Ch. 82.30 km to 112.5 km	434.71	148.8	413.74
KBC Ch. 133.519 km to 189.977 km	375.74	107.63	497.12
KBC Pumping Station No. 1	187.3	Pumps and Civil Work	
KBC Pumping Station No. 2	187.3	Pumps and Civil Work	
KBC Pumping Station No. 3	141.2	Pumps and Civil Work	

Wherever possible, works of distributaries and minors are also being taken up concurrently. With this strategy a command area of about 48,000 ha out of total command of 366,000 ha of KBC is being covered.

Tenders for Vejpur, Malsan and Madka branch canals are already awarded and work is in progress. In Banaskantha area, EPC packages worth over Rs. 2 billion have been invited for two Branch Canals, Dhima and Gadsisar, and their distribution network. Item rate tenders worth around Rs. 3 billion have been invited for the distribution network of Radhanpur, Manpura, Vejpur and Malsan branch canals. Thus works have been taken up to develop a command area of about 175,000 ha in north Gujarat.

Such combination of Item Rate and EPC contracts, each with a time limit of 18 to 24 months, will help in early realisation of the potential benefits of SSP in the command areas in Patan, Banaskantha, Kachchh and Saurashtra regions.

Optimum Utilisation of Working Seasons

A detailed programme was worked out for inviting tenders for the remaining canal works within three months so that all the works can begin at the earliest and completed soon.

State-of-the-art tools for Construction Project Management are being used to prepare a meticulously prepared project-specific Work Plan taking into account all resources and constraints. Project Engineers are being trained in this subject to enable them to monitor the progress of each component.

Project Management Consultants

Each construction package of the canal network is a project by itself and has to be monitored and managed professionally. Project Monitoring Consultants are, therefore, being appointed to oversee execution so as to cope with technical manpower shortage and to effectively and efficiently monitor the Project for timely completion.

Third Party Quality Audit Consultant

Each contractor is primarily responsible and contractually obliged for quality assurance of work. Operation and maintenance for a further period of two to five years is also an integral part of the contract. However, to further ensure and monitor the quality of works on sustained basis, a third party quality consultant is being appointed to monitor and ensure the quality of the works in progress.

Participatory Irrigation Management

For equitable distribution and efficient utilisation of water in the command areas, water for irrigation is planned to be delivered through Water Users' Association—a co-operative society of farmers—through the provisions of Participatory Irrigation Management (PIM) Act, 2007. The WUAs are also expected to operate and maintain the delivery system including the Minors and Sub-minors, and recover water charges from users. This would ensure better physical performance and financial working of the irrigation system. A total of 4,780 WUAs are to be formed in the command area in 14 districts, of which about 1,600 have already been registered. Efforts to sensitise and operationalise more WUAs are being made, using methods ranging from interactive group discussions to intensive training programmes.

Multidisciplinary Coordination Committee

A district level committee, headed by the Collector, has been constituted for up-scaling PIM activities. This Committee consists of elected members, representatives of WUAs, Deputy Conservator of Forest, District Agriculture Officer, Engineer-in-Charge of command area. It also includes NGOs, chosen to promote WUAs to elicit farmer participation in water distribution, and to facilitate efficient water use and adoption of best water use practices, encourage PIM, resolve local issues effectively and in a timely manner. This committee has also been delegated financial powers to deal with various issues. The initiative would also help sensitise and operationalise the WUAs effectively.

Separate Wing for O&M of Main Canal

NMC has been completed and the water has been flowing right up to Rajasthan since March 2008. The 458 km long main canal, which is 458 km long, is the artery created under SSP to transport water to different regions. To operate and maintain this canal comprising more than 600 structures, a separate wing has been created and made functional through three divisional offices equipped with necessary manpower and infrastructure.

Promoting Micro-Irrigation

Under the Command area, in order to popularise MIS, a pilot project each is being implemented in 73 talukas. This is proposed dovetail the State's scheme with the Central Government Scheme under CADWM in order to facilitate larger fund flow. GoG has also taken pioneering initiatives to promote Pressurised Irrigation Network System (PINS) and MIS. The Project Authorities have undertaken the exercise of exploring the potential of implementing drip irrigation practices wherever possible, considering the existing cropping pattern and its suitability for drip technology. The State government has set up a Special Purpose Vehicle to implement it, called Gujarat Green Revolution Company (GGRC). The government is also providing a subsidy to the extent of 50 per cent for drip irrigation in the normal areas and 75 per cent in the tribal areas towards capital costs. Project Authorities and GGRC are taking all steps to popularise this scheme in the command area. The initiative of the government to introduce metering and *pro-rata* pricing of electricity for new agricultural power connections is encouraging farmers to adopt micro-irrigation systems.

Financial Resources

With the Planning Commission's nod to the revised estimate of Rs. 392.4 billion for the project and the Environmental Sub-group of NCA's consent to raise the dam height to 138.68 metres, the project is all set to embark upon the remaining works in the next three financial years and to create additional 1.3 MHa Irrigation Potential. Apart from enhanced state provisions in terms of Annual Development Plans, financial support under Accelerated Irrigation Benefit Programme (AIBP) is also being sought. A detailed region-wise and year-wise physical and financial plan has been made to enable this. The project's eligibilities under AIBP are given below:

Year	*Proposed IP Creation (up to minor) Ha.*	*Eligible AIBP grant Rs. Crore*
2010-11	1,06,363	1590
2011-12	2,50008	3217
2012-13	5,31,579	2384
2012-14	4,08,933	1178
Total	**12,96,883**	**8279**

E-Tendering

All major bids are invited in e-tendering mode so as to ensure transparency and responsiveness in the process. Competitive bidding in two stages—Technical and Financial—is followed to arrive at the least-price workable contracts. Tenders worth about Rs. 50 billion have been invited during the financial year 2010-11. Of these, tenders for 51 works worth around Rs. 10.97 billion have been evaluated and finalised and for another 49 works worth around Rs 39.1 billion tender evaluation is in process.

Rehabilitation with a Human Face

SSP has an R&R Policy with a human face. As noted by many observers, it is one of the best in the world (Patel, 2001; Fisher, 2001). The policy of "land for land", and a minimum of 2 ha of cultivable land to each adult member of a PAF, apart from all basic amenities in the village of rehabilitation, had not been followed anywhere in independent India's dam building history. Resettlement work is substantially over. The Gujarat PAFs have been given land as per their choice and other benefits. Some 4,750 PAFs are resettled at 110 R&R sites. PAFs of MP willing to resettle in Gujarat are given written offers with choices. Special efforts are made to bring them to Gujarat for land selection. GoG has created a Land Bank for PAFs from MP to facilitate the selection and allotment process, and 5,588 PAFs are resettled in 108 R&R sites in Gujarat. From Maharashtra, 769 PAFs have been resettled in 18 R&R sites in Gujarat. The R&R Sub-group of NCA monitors the progress made by Sardar Sarovar Punarvasvat Agency (SSPA). Monitoring and Evaluation studies have been entrusted to independent social science agencies. The Centre for Social Studies (CSS), Surat—the monitoring agency for the Gujarat—has prepared monographs on each of the submerging villages located in Gujarat, highlighting their socio-economic and cultural status before resettlement. These studies show that there is least difficulty in resettlement and rehabilitation of the tribal population, who have welcomed the opportunity to join the mainstream of socio-economic development. "To insist that people must neither change nor learn non traditional skills is putting a premium on preciousity. It is no less than asking the tribal people to remain in a museum, to become fossilized. This is a demand that outrages all human values, all democratic rights. It can, in fact, end up destroying them just as surely as forcible displacement will." (Fisher, *et al.*, 1997: 286). And rightly so, because they also have the Right to Development, Right to Participation, Right to Work, Right to Education, Right to Health, and all the other rights that every Indian citizen is entitled to.

Environmental Sustainability

Sustainable environmental and ecological measures are implemented as a part of project activities and they are regularly monitored for sustained effectiveness. Concurrently with the efforts of early completion of remaining physical infrastructure works, measures to safeguard the command area environment from adverse impacts are also being taken. An Environment Safeguard Measures Plan has been prepared and is being implemented for Command Area Phase-I which has been appreciated by GoI. Moreover, to conserve the biodiversity of upstream region as well as command area, a Biodiversity plan is implemented by Forest Department at project cost, with a view to preserve biodiversity of the upstream region. MSU also has been enlisted in the efforts for *in-situ* conservation of the flora in the command area.

The road map for getting environmental clearance for the final stage of raising the dam height has been put into action. Implementation of the action plan of Shoolpaneshwar Sanctuary has been initiated and is planned to be completed in the next two years. To plan short-and long-term remedial measures for Downstream Management, a study for monitoring the water quality of the reservoir by Central Water and Power Research Station (CWPRS), Pune and another for downstream ecological impacts by Central Inland Fisheries Research Institute (CIFRI), Barrackpore have been taken up. Morphological changes are also being studied by CWPRS. Wildlife Clearance to KBC became a challenge as it passes through the Wild Ass Sanctuary, the habitat for one of the rare species. Substantial provisions of and additional Rs. 3 billion have been made for providing safe passage to the animals across the canal through extra large structures.

All these efforts to conserve the environment will supplement the benefits from irrigation, which will lead to sustainable socio-economic development of the area.

Conjunctive Use of Water for Irrigation

Against the planned use of 2.7 MAF groundwater annually the present use is about 0.8 MAF only, leaving a lot of scope for planned implementation of the conjunctive use strategy in the SSP command. To mitigate any threat of waterlogging, thereby protecting environment, effective planning for drainage is essential. Therefore, completed drainage studies have been transformed into an action plan. The conjunctive use of water is envisaged to check the rising water table. In view of this a pilot project for conjunctive use has also been initiated in the Phase-I area. A further plan for Phase-II is also under preparation.

Separate Feeder Pipelines for Drinking Water Security

Narmada based drinking water supply schemes are supplying water to almost half of Gujarat state, even in areas outside the irrigation command. The drinking water supply to Panchmahal, Sabarkantha, Mehsana, Patan and Banaskantha districts is through the Narmada Main Canal. As the canals meant for irrigation are also required to cater to drinking and domestic water demands, their closure for preventive maintenance becomes difficult. To minimise conflicts of interest between farmers and drinking water beneficiaries, approval of the State Government has been obtained for separate feeder pipelines for drinking water supply. This would also help avoid operating irrigation canals like the Saurashtra Branch and its Vallabhipur and Maliya sub-brances throughout the year. Separate provision of feeder pipelines would enable project engineers to undertake periodic maintenance.

Highly reliable water supply to 6,513 villages and 114 towns is ensured in water scarce areas. The extension of water supply to newer areas is in progress which would result in round-the-year supply of drinking water to 9,633 villages and 131 towns/cities in the water stressed regions of the state. Needless to say, clean and potable water supply from SSP to the water-stressed part of Saurashtra, with a high incidence of fluoride and nitrate in its ground water, has had a cascading effect on various aspects of the lives of the local people. Water supply through tankers has rapidly become history here.

Thus, to ensure reliable and assured supply of water, the State Government has decided to create a separate drinking water transmission and distribution system at a cost of about Rs. 44 billion. The Water Supply department has received in-principle budgetary approval from the government for its implementation. This initiative would help avoid conflict between the supply of irrigation water and drinking water; enable the Project Authorities necessary closures of canals for maintenance besides enhancing the reliability of supply of clean drinking water to a large section of the State and thereby ensuring the water security.

Prompt Decisions on Technical Issues

The project, in its course of implementation, meets with several technical challenges, some of which are of great magnitude. Taking prompt decisions on such technical issues is of vital importance for speedy implementation of the project activities. A Board of Consultants comprising eminent engineers namely Shri R. Jeyaseelan, Shri M.U. Purohit, Dr. A.K. Dhawan, Dr. A.V. Shroff and Shri Kannaiah Naidu has been constituted to advise the project engineers.

Initiating a Pilot Project on Canal Automation

All the canal regulating structures of SSP with discharge carrying capacity of more than 8.5 cumecs are to be operated automatically with a view to ensure safe, flexible and efficient operation. The Pilot Project of Canal Automation, a SCADA based Remote Monitoring and Control System, has been revived and its scope has been extended to include the entire Narmada Main Canal besides the Vadodara and Saurashtra branch canals. Thus the total network covered under the Pilot Project would be about 600 km in length, involving about 200 structures with more than 600 gates. Telecommunications Consultants India Limited (TCIL) has been engaged to review and finalise the Bid Documents for this Pilot Project and its tenders will be invited in the first quarter of the financial year 2011-12.

The "End of Thirst"

The last decade of the 'with SSP' scenario has shown how the role of Narmada water is going to be larger than planned, hence the urgent need to complete the remaining works. It may be worth reiterating that the State Government and the Project Authorities are committed to complete SSP at the earliest as planned, and deliver its full benefits to all stakeholders by providing much needed water security for irrigation, drinking and industrial use. The Project Authorities are also open to implement new ideas/technologies that enhance project efficiency and efficacy without compromising on its basic objective of providing water for irrigation in its command area as per the plan.

References

Acharya, S., S.P. Punalekar, N.K. Gourah and N.V.V. Char, (1992): *Report of Fact-finding Team Appointed by Ministry of Water Resources on Manibeli*, Government of India. New Delhi.

Ahluwalia, M.S. (2000): "Economic Performance of States in Post-Reforms Period," *Economic and Political Weekly*, May 6-12.

Ahmedabad Municipal Corporation (AMC) (2010): *Information on Reforms Achieved for Consideration under The Awards for Excellence In Municipal Reforms and Service Delivery Improvements*, Report Submitted to Ministry of Urban Development (MoUD), Government of India.

Alagh, Y.K. (2010): "A Sardar Sarovar riddle", *The Financial Express*, Ahmedabad Edition, April 19.

Alagh, Y.K., Mahesh Pathak, and D.T. Buch (1995): *Narmada and Environment—An Assessment.* Har Nand Publications. Delhi.

Alagh, Y.K., R.D. Desai, G.S. Guha, and S.P. Kashyap (1995): *Economic Dimensions of the Sardar Sarovar Project.* Har Anand Publications. SPIESR. Ahmedabad.

Alexander, K.C., R.R. Prasad and M.P. Jahagirdar (1991): *Tribals, Rehabilitation and Development,* Rawat Publications. Jaipur.

Allen, R.G., L.S. Willardson and H. Frederiksen (1998): "Water Use Definitions and Their Use for Assessing the Impacts of Water Conservation" in J.M. de Jager; L.P.Vermes and R. Rageb (eds.), Proceedings ICID Workshop on *Sustainable Irrigation in Areas of Water Scarcity and Drought*, Oxford, England. September, 11-12.

Alvares, C. and R. Billorey (1988): *Damming the Narmada.* Natraj Publishers. Dehra Dun.

Alvares, Claude and Ramesh Billorey (1987): "Damming the Narmada: The Politics Behind the Destruction," *The Ecologist*, Vol. 17, No. 2.

— (1988): *Damming the Narmada: India's Greatest Planned Environmental Disaster,* Natraj Publishers. Dehradun.

— (1988): *Damming the Narmada,* Third World Network. Penang, Malaysia.

Amarasinghe, Upali *et al.* (2005): *Spatial Variation in Water Supply and Demand across River Basins of India,* Research Report; 83, International Water Management Institute, Colombo.

ARCH-Vahini (1988a): *Displacement in Sardar Sarovar (Narmada) Project: A Gujarat Experience*, Mangrol, Rajapipla. April.

— (1988b): *Sardar Sarovar Oustees – Which Way to Go: An Activist's Dilemma*, Mangrol, Rajpipla. September.

— (1991): "Resettlement and Rehabilitation in SSP: Are the Critics Right?" (*Mimeo*).

— (1992a): *Is R&R Impossible in SSP: The Role of NBA (India), EDF (USA) and MOEF (India): A Submission to Morse Mission.* Mangrol, Rajpipla. February.

— (1992b): *A Critique of Memorandum Submitted to the World Bank by the EDF and Other NGOs* on *March 20, 1991*, Mangrol, Rajpipla. September.

— (1992c): *Letter to Mr. Lewis Preston, President, World Bank on Morse Report,* Mangrol, Rajapipla.

— (1992d): *Letter to David Jones, Associate Director, Oxfam U.K. on Morse Report*, Mangrol, Rajpipla. September.

— (1992e): *The Independent Review: A Search for Truth?* Mangrol, Rajpipla. November.

— (1993a): *Morse-Berger Review of Sardar Sarovar Project: A Counterview*, Mangrol, Rajpipla. January.

— (1993b): *The SSP Withdrawn from the World Bank: ARCH-Vahini's Views*, Mangrol, Rajpipla. April.

— (1993c): *The Sardar Sarovar Project: An Intellectual Fashion*, Mangrol, Rajpipla. July.

— (1993d): *Human Rights in the Sardar Sarovar Project: A Submission to Fact Finding Committee of PUCL,* Mangrol, Rajpipla. October.

— (1994): *A Note of PAPs in M.P. who are Entitled for Agricultural Land for Research,* Mangrol, Rajpipla. March.

— (1995): *R&R in SSP: Are the Critics Right?* in Pathak, Mahesh (ed.) *Development Imperatives and Environmental Concerns.* Agro Economic Research Centre. Vallabh Vidyanagar.

ASCE (1987): *Planning, Operation, Rehabilitation and Automation of Irrigation Water Delivery System,* Proceeding of ASCE Symposium, Portland, Oregon.

— (1993): "Special Issue of Canal System Modelling", *Journal of Irrigation and Drainage,* No. 119 (4).

Barot, J.M. (1996): *Occurrence of Salinity in Gujurat in Reaching the Unreached: Challenges for the 21st Century*, 22nd WEDC Conference, New Delhi, India.

Baviskar, Amita (1991): "The Researcher as a Pilgrim," *Lokayan Bulletin*, Volume 9 (3-4). March.

Berg, Elliot and Mary M. Shirley (1987): *Divestiture in Developing Countries,* World Bank Discussion Paper No. 11, Washington, DC.

Bhaskaracharya Institute for Space Applications and Geo-informatics (2009): *Geographical Information System for Sardar Sarovar Project (SSP) Command Area Phase I: Project Proposal*, Gandhinagar.

Bhatia, Bela (1992): "Lush Fields and Parched Throats – Political Economy of Groundwater in Gujarat", *Economic and Political Weekly*, December 19-26.

— (1997): "Forced Eviction in Narmada Valley" in Jean Dreze; M. Samson and S. Singh (eds.) *The Dam and the Nation,* Oxford University Press. Delhi.

Border Environment Cooperation Commission (2003): "Hidalgo County Irrigation District

Brewer, J., S. Kolavalli; A. H. Kalro; G. Naik; S. Ramnarayan; K.V. Raju; and R. Sakthivadivel (1999): *Irrigation Management Transfer in India: Policies, Processes, and Performance,* A collaborative study by the Indian Institute of Management, Ahmedabad, India and the International Irrigation Management Institute, Colombo, Sri Lanka. Oxford & IBH Publishing Co. Pvt. Ltd, New Delhi.

Bribiesca, Sanchez (1967): *9th International Commission on Large Dams.* Turkey.

Bruns, Bryan Randolph and Ruth Meinzen-Dick (2000): *Negotiating Water Rights,* Sage, New Delhi.

Bureau of Indian Standards (1986): *Criteria for Earthquake-Resistant Design of Structures* in *Indian Standard IS 1893: 1984.* New Delhi.

CAG Report (2000 and 2001): Comptroller and Auditor General Report 2002, Government of India.

Census of India (2001): *Housing and Amenities Table, Gujarat.*

Central Board of Irrigation and Power (1993): *Canal Automation*, Publication No. 238, New Delhi.

Central inland Capture Fisheries Research Institute (1993): *Studies on Fish Conservation in Narmada Sagar, Sardar Sarovar and it's Downstream,* Report submitted to Narmada Control Authority (NCA), Indore, Government of India.

Central Soil and Water Conservation Research and Training Institute (n.d.): Interim Report on *Rainfall Runoff and Sediment Yield Measurement in Catchment Area of Sardar Sarovar Project lying within Gujarat.* Vasad.

CEPT - CRISIL (2006): *Public Private Partnership: A Study of Potential Benefits in the Sardar Sarovar Narmada Project* (Unpublished), Submitted to Sardar Sarovar Narmada Nigam Limited, Gandhinagar.

Cernea, Michael (1995): *Keynote Address: Understanding and Preventing Impoverishment from Displacement* in *International Conference on Development Induced Impoverishment.* University of Oxford. Oxford.

— (1991): "Involving Settlement: Social Research, Policy and Planning" in Michael M. Cernea (ed.) *Putting People First: Sociological Variables in Rural Development*, Oxford University Press, Second Edition. New York.

— (1996): "Public Policy Responses to Development-induced Population Displacements," *Economic and Political Weekly*, Vol. 31 (24).

Chakravorty, Ujjayant and Chieko Umetsu (2003): "Basinwide Water Management: A Spatial Model", *Journal of Environmental Economics and Management*, 45 (2003): 1-23.

Chambers, Robert (1988): *Managing Canal Irrigation: Practical Analysis from South Asia,* Oxford and IBH, New Delhi, India.

Chandra, A. and C. Sudhir (2010): *A Study of Kolhapur Lift Irrigation Co-operatives*, MTS Report, Institute of Rural Management, Anand.

Choudhury, N. and V. Kher (2006): "Public-Private Partnership in Surface Water Irrigation: A Case of Kolhapur", IWMI Tata Water Policy Program, Anand (unpublished research).

CICFRI (n.d.): *Studies on fish Conservation in Narmada Sagar, Sardar Sarovar and its Downstream.* Barackpore.

CMIE (2010): *India Harvest Database*, Centre for Monitoring Indian Economy.

Commissioner of Fisheries, Govt. of Maharashtra's letter No. Fish/T/011403/2168 to Secretary, Narmada Ghati Development Department, Government of MP.

Commissioner of Health, Medical Services and Education (1993): *Environmental Impact Assessment Studies on water Related Diseases in SSP command Area Including the Area Downstream of the Narmada Dam.* Government of Gujarat. Gandhinagar.

Cook, Paul and Colin Kirkpatrick (eds.) (1988): *Privatization in Less Developed Countries,* St.Martin's Press, New York.

Cooter, Robert D. (1997): "The Rule of State Law and the Rule-of-law State: The Economic Analysis of the Legal Foundations of Development" in Micheal Bruno and Boris Pleskovic (eds.) *Annual World Bank Conference on Development Economics 1996*, World Bank, Washington, D.C.

Cowan, L. Gray (1990): *Privatization in the Developing World,* Greenwood Press, New York.

Cullet, Philippe (2007): *The Sardar Sarovar Dam Project: Selected Documents*, Ashgate Publishing Company. Hampshire. U.K.

Dalua, A.K. (1993): *Environmental Impact of Large Reservoir Projects on Human Settlement*, Ashish Publishing House. New Delhi.

Dam Safety Panel (1997): *Report of the Thirty Fourth Meeting of Dam Safety Panel, Sardar Sarovar (Narmada) Project,.* Sardar Sarovar Narmada Nigam Limited. *Gujarat.*

DANIDA (1996): *Community Participation in Rural Drinking Water and Sanitation Projects: A Comparative Study of Five Projects in India*, RWS & Sanitation Group, SA, UNDP-WB & IIMA.

Das, Biswaroop (1983): "Studies on Rehabilitation of Submerging Villages – Vadgam" *(Mimeo).* Centre for Social Studies. Surat.

— (1986): "Submerging Tribal Belt in Gujarat: Spatial Variation and Stratification," *State and Society*. 6(1).

— and A.S. Charan (1983): "Cost of Submergence," *(Mimeo).* Centre for Social Studies. Surat.

Datta, Samar K. (ed) (2003): *Cooperatives in Agriculture – A Millennium Study* (Volume 24), Academic Foundation, Delhi.

— *et al* (forthcoming): *Developing India's Strategies Response to the Global Debate over Fisheries Subsidies*, Chapter 4, Allied Publishers, Delhi.

Daud Committe (2001): *Report of the Committee to Assist the Resettlement and Rehabilitation of the Sardar Sarovar Project-Affected Persons*, Government of Maharashtra.

Dave Kapil (2010): "Half of Gujarat will live in cities". *Daily News and Analysis*, June, 2.

Davies, Susanna and Naomi Hossain (1997): "*Livelihood Adaptation, Public Action and Civil Society," Working Paper 57*, IDS, Brighton.

De Fraiture, Charlotte and Chris Perry (2002): "Why is irrigation water demand inelastic at low price ranges?" Paper presented at the conference on *Irrigation Water Policies: Micro and Macro Considerations*, Agadir, Morocco, June, 15-17.

Department of Botany (1983): *The Sardar Sarovar Narmada Project Studies on Ecology and Environment.* M.S. University. Baroda.

— (1992): *Eco Environmental and Wildlife Management Studies on SSP Submergence Area in Gujarat.* M.S. University. Baroda.

Department of Earthquake Engineering (1982): *Design Earthquake Parameters for Navagam Dam Site in Gujarat.* University of Roorkee.

Department of Environment and Forests (1986): *Environmental Aspects of Narmada Sagar and Sardar Sarovar Multi-purpose Project*, URL: http://www.ielrc.org/content/c8601.pdf. December.

Desai Vaidya, Prerana (2010): "Impact Of Sardar Sarovar Project: A Field Evaluation." *Seminar on Evolution and Growth of Sardar Sarovar, the Lifeline of Gujarat,* CEPT University, Ahmedabad.

Desai, G.M., Gurdev Singh and D.C. Sah (1979): "Impact of Scarcity on Farm Economy and Significance of Relief Operations," *(Mimeo).* Indian Institute of Management. Ahmedabad.

Desai, N.B. (1990): *"Influence of Sedimentation on Service Life of Sardar Sarovar," Indian Journal and Power and River Valley Development, Special Number on Sardar Sarovar Narmada Project,* September (pp. 249-253)

Desai, S.J. and M.B. Joshi (2008): *Narmada Water Plays Its Role – Capturing Initial Trends from Gujarat,* Presentation, Sardar Sarovar Narmada Nigam Ltd., Gandhinagar.

Deshpande, N. (2009): *Displaced by the Dam: Living on the Edge,* URL (http://www.indiatogether.org/2009/oct/hrt-sardar.htm), October.

Dharmadhikary, S. (2005): *Unravelling Bhakhra: Assessing the Temple of Resurgent India*, Manthan Adhyayan Kendra, Bhopal.

Dhawan, B.D. (ed.) (1990): *Big Dams: Claims, Counter Claims,* Commonwealth Publishers, New Delhi.

Dholakia, Ravindra H. (1985): *Regional Disparity in Economic Growth in India.* Himalaya Publishing House. Bombay.

—(1994): *Spatial Dimensions of Acceleration in Economic Growth in India* in Economic and Political Weekly. Volume 29, Issue 35 (pp. 2303-09)

—(2010): *Has Agriculture in Gujarat Shifted to High Growth Path?* in Dholakia, Ravindra H. and Samar K. Datta (eds.): *High Growth Trajectory and Structural Changes in Gujarat Agriculture*, Macmillan Publishers.

—and Samar K. Datta (eds.) (2010): *High Growth Trajectory and Structural Changes in Gujarat Agriculture,* Macmillan Publishers.

—and Ajay N. Oza (1999): *Microeconomics for Management Students*, 2nd Edition. Oxford University Press. Delhi.

Dixon J.A., L.M. Talbot and G.J.M. Le Moigne (1989): *Dams and the Environment: Considerations in World Bank Projects*, World Bank Technical Paper Number 110, Washington D.C.

Domanski, Tomasz (1992): "Development of Small Private Companies and Their Marketing Activities," in *Journal of Business Research,* Vol. 24, pp. 57-65.

Falkenmark, Malin and Carl Widstrand (1992): "Population and Water Resources: A Delicate Balance." *Population Bulletin,* Population Reference Bureau: Washington, DC.

FAO (2008): *Report of the Lake Victoria Fisheries Organization and FAO - National Stakeholders' Workshops on Fishing Effort and Capacity on Lake Victoria (2006),* FAO Fisheries and Aquaculture Report No. 817, Food and Agriculture Organization of the United Nations, Rome.

Fernandez, A.P. (1990): "No Dam is an Island" in A P Fernandez (ed.) *Workshop on Rehabilitation of Persons Displaced by Development Projects: A Report*, Institute for Social and Economic Change and MYRADA. Bangalore.

Final Order and Decision of the Narmada Water Dispute Tribunal (NWDT) Gazetted on December 12, 1979, Source: nca.gov.in/forms_pdf/ nwdt_finalorder.pdf

Fisher, William F. (2001): "Diverting the Water: Revisiting the Sardar Sarovar Project", *International Journal of Water Resources Development*, 17 (3): 303-314.

Fitch, Lyle C. (1988): "The Rocky Road to Privatization," in *American Journal of Economics and Sociology*, Vol.47, No. 1, pp. 1-14.

Frederick, K D (1992): *Balancing Water Demand with Supplies: The Role of Management in a World of Increasing Scarcity*, Technical Paper 189, Washington DC: World Bank.

GoI (1987): *National Water Policy: 1987,* Government of India, Ministry of Water Resources. New Delhi.

GoI (2002): *National Water Policy: 2002*, Government of India, Ministry of Water Resources, New Delhi.

Goldsmith, Edward, and Nicholas Hildyard (1984): *The Social and Environmental Effects of Large Dams,* Sierra Club Books. San Francisco.

GOPA Consultants (1991): *Narmada River Basin Development Project –* Report submitted to SSSNL, Government of Gujarat. http:// www.nvda.nic.in/env.htm

Government of Alberta (2003): "Alberta's irrigation infrastructure," http:// www1.agric.gov.ab.ca/$department/deptdocs.nsf/all/irr7197/$FILE/ irrigationinalta-part2.pdf (accessed on 11th November, 2009).

Government of Gujarat (1987): *Compilation of Government Resolutions for Rehabilitation and Resettlement Policy for Oustees of Sardar Sarovar Project*, Narmada Development Department. Gandhinagar.

— (1992): *No Cancelling Tomorrow: A Critique of the Report of the Independence Mission of Gujarat*, Gandhinagar.

— (1992): *Comment on the Report of the Independent Review Mission on SSP* Gandhinagar.

— (1997): *Water Supply & Sanitation, 9th Five-Year Plan (1997-2002)*, Narmada Water Resource & Water Supply Department, Government of Gujarat, Gandhinagar.

— (1998): *Infrastructure 2010: Master Plan on Drinking Water Supply*, Narmada Water Resources and Water Supply Department, Gandhinagar.

— (2000): *Drinking Water Scarcity 2000-2001*, Narmada Water Resources & Water Supply Department, Gandhinagar.

— (2002): *10th Five-Year Plan (2002-2007): Water Supply & Sanitation*, Narmada Water Resource & Water Supply Department, Government of Gujarat, Gandhinagar.

— (2007): *Gujarat Water Users' Participatory Irrigation Management Act*, Government of Gujarat, Gandhinagar.

— (2009): *First Approximation report to increase groundwater resources through artificial recharge in Gujarat State*, Narmada Water Resources, Water Supply and Kalpsar Department, Gandhinagar.

— (2010): *Report of the Expert Group for Strategy for the Accelerated Development of Sardar Sarovar Project Command Area,* Sardar Sarovar Narmada Nigam Ltd., Gandhinagar (unpublished report).

Government of India (1978): *The Report of the Narmada Water Dispute Tribunal with its Decision*, Vols. I and II. New Delhi.

— (1990): *Approach to Eighth Five Year Plan*, Planning Commission. New Delhi.

— (1992a): *Eighth Five Year Plan (1992-97), Vol. II: Sectoral Programmes of Development*, Planning Commission. New Delhi.

— (1992b): *Report of the Five Member Group Set up by the Ministry of Water Resources to Discuss Various Issues Relating to the Sardar Sarovar Project*, Vols. I and II. New Delhi.

— (1987): *National Water Policy, 1987*, Government of India, Ministry of Water Resources.

— (2002): *National Water Policy, 2002, Government of India, Ministry of Water Resources.*

— (1995): *Report of the Committee on Private Sector Participation in Major and Medium Irrigation Projects*, Ministry of Water Resources, New Delhi, India.

— (2002): *National Water Policy,* Ministry of Water Resources, New Delhi, India.

— (2005): *Report on 3rd Census of Minor Irrigation Schemes*, Ministry of Water Resources, Minor Irrigation Division, New Delhi.

Government of Madhya Pradesh (1992): *Action Plan of R&R of Oustees of MP (SSP), Parts I and II,* Narmada Valley Development Authority.

Guggenheim, Scott (1990): "Development and Dynamics of Displacement" in A.P. Fernandez (ed.) *Workshop on Rehabilitation of Persons Displaced by Development Projects: A Report*, Institute for Social and Economic Change and MYRADA. Bangalore.

Gujarat Infrastructure Development Board (2009): *BIG 2020 – Blueprint for Infrastructure in Gujarat*, Gandhinagar.

Gujarat PUCL (1992): Human Rights Issues at Manibeli: Fictions and Facts *(Mimeo),* Report by Gujarat PUCL on behalf of Human Rights Group of Narmada Abhiyan.

— (1993), "Drowning of Manibeli," *PUCL Bulletin*, Vol. 12(7), Report by M. Mohanty, C.C. Kalara on behalf of Human Rights Campaign on Manibeli.

Gulati, Ashok; Tushaar Shah and Ganga Shreedhar (2009): *Agriculture Performance in Gujarat since 2000: Can it be a* Divadandi *(Lighthouse) for other States?* International Water Management Institute, Anand; and International Food Policy Research Institute, New Delhi.

Gupta, Rajiv K (2001): "Human Rights Dimension of Regional Water Transfer: Experience of the Sardar Sarovar Project," *International Journal of Water Resources Development*, Vol. 17 (1).

GWRDC (n.d.): *Ground Water Resource Estimation*. Retrieved from http://www.gwrdc.gujarat.gov.in/Gwremaps/GWRE_B.K..htm

GWSSB (1983): *Water Demand for Non-agriculture Use from Narmada Project*. Gandhinagar.

GWSSB (1985-2005): *Master Plans*, Gujarat Water Supply and Sewerage Board, Gandhinagar.

— (2003): *Sardar Sarovar Canal Based Drinking Water Supply Master Plan*, Gujarat Water Supply & Sewerage Board, Gandhinagar.

Hague Deceleration (2000): "Ministerial Declaration of The Hague on Water Security in the 21st Century," *Second World Water Forum*, Hague. March.

Hanke, S.H. (ed.) (1987): *Privatization and Development*, Institute for Contemporary Studies Press, San Francisco.

Haskoning (1999): *Netherlands Support to the Water Sector in India: Impacts on Institutional and Development and Change*, Haskoning, Netherlands.

Hazare, Anna (1997): *Ralegaon Siddhi: A Veritable Transformation translated by Pendse*, S.B. from original Marathi book, published by Ralegaon Siddhi Pariwar.

Heath, John (1989): "Privatization: Modalities and Strategy," in V.V. Ramanadham (ed.) *Privatization in Developing Countries*, Routledge, London, pp. 118-124.

Hirway, I (2005): "Ensuring Drinking Water to All: A Study in Gujarat." *4th IWMI-TATA Annual Partners Research Meet*, February 24-26.

— (1999), *Integrating Environmental & Economic Accounting: The Case of Gujarat – Phase 1*, Gujarat Ecology Commission, Vadodara, Prepared by Centre for Development Alternatives, Ahmedabad.

— and Shital Lodhia (2004): *Status of Drinking Water in Gujarat: Towards a Sustainable Approach*, Unpublished Report submitted to World Health Organisation, New Delhi.

— and P.P. Patel (1994): *Dynamics of Drinking Water in Rural Gujarat*, Action Aid India, Bangalore.

— and Subhrangsu Goswami (2008): "Functioning of the Drinking Water Component of the Narmada Pipeline Project in Gujarat," Special Article, *Economic and Political Weekly*, March 2008.

— *et al.* (2002): *Quick Valuation of Depletion and Degradation of Environmental Resources in Gujarat*, sponsored by Gujarat Ecology Commission, Vadodara, Prepared by Centre For Development Alternatives, Ahmedabad.

— and Subhrangsu Goswami (2005): "Performance of Narmada Pipeline Project in Gujarat", *Research Highlights No. 10*, International Water Management Institute, Anand.

— (2006 a): *Concurrent Monitoring of Worlds Largest Pipeline Drinking Water Project: Study of Narmada Pipeline Project in Gujarat*, Unpublished Draft Report submitted to International Water Management Institute, Anand and PRAVAH, Anand.

— (2006 b): *Status of Drinking Water in Coastal Saurashtra: Need for Area Specific Approach*, Unpublished Draft Report submitted to Coastal Salinity Cell, Rajkot.

— (2008): "Functioning of the Drinking Water Component of the Narmada Pipeline Project in Gujarat", Special Article, *Economic and Political Weekly*, March.

Hooja, Rakesh; Ganesh Pangare; and K.V. Raju (2002): *Users in Water Management: The Andhra Model and Its Replicability in India,* Rawat Publications. New Delhi, India.

Howe, Charles W, Dennis R. Schurmeier and W. Douglas Shaw, Jr. (1986) "Innovative Approaches to Water Allocation: The Potential for Water Markets," *Water Resources Research*, 22 (4), pp. 439-445, April.

Independent Review Mission (1991): *Independent Review, Sardar Sarovar.* World Bank.

Indian Express (2009): "Government orders removal of diesel pumps installed along Narmada canal.", Ahmedabad edition, June 20.

Indu, Rajnarayan and Alka Rawal (2007): "Impact of High Incidence of Kidney Stone: A Study of Coastal Villages in Junagadh, Gujarat", Paper presented in workshop on *Crisis in Drinking Water in Coastal India*, EDI, Ahmedabad.

Institute of Rural Management, Anand (IRMA) (2001): *White Paper on Water in Gujarat*, report prepared for Government of Gujarat, IRMA and UNICEF, Gandhinagar.

International Council of Scientific Unions (1972): *Man Made Lakes as Modified Ecosystem.* Scope Report 2. Paris.

International Water Power and Dam Construction (1988): *Reservior and the Environment—Experience in Managment and Monitoring*. Volume 40, Issue 9.

IPCC (Intergovernmental Panel on Climate Change) (2001): *Impacts, Adaptation, and Vulnerability*. Cambridge, England, Cambridge University Press.

IRMA – Institute of Rural Management (2001): *White Paper on Water in Gujarat*, prepared for the Department of Narmada, Water Resources and Water Supply, Government of Gujarat, Gandhinagar.

Iyengar, Sudarshan (1997): *Resettlement and Rehabilitation in SSP: A Review of Policy and Implementation.* GIDR. Ahmedabad.

—, and Vinayak Dave (1998): *Status of Project-affected Families from Madhya Pradesh Resettled in Gujarat.* Reconnaissance Report. GIDR. Ahmedabad.

Iyengar, Sudarshan (1998): *Gujarat* in V.M. Rao (ed.) *The Poor in a Hostile Society: Glimpses of Changing Poverty Scenario in India*, Vikas Publishing House Pvt. Ltd., New Delhi.

Jayaratne, R.S. (1990): "Involuntary Resettlement Consequent to the Construction of Three Large Dams and Reservoirs under the Mahaweli Development Project, Sri Lanka" in A.P. Fernandez (ed.) *Workshop on Rehabilitation of Persons Displaced by Development Projects: A Report*, Institute for Social and Economic Change and MYRADA. Bangalore.

Jodha, N.S. (1981): "Role of Credit in Farmers' Adjustment against Risk in Arid and Semiarid Tropical Areas of India," *Economic and Political Weekly*, Vol. 26 (42-43).

Johnson, III Sam H. (1997): "Irrigation Management Transfer: Decentrlizing Public Irrigation in Mexico," in *Water International*, 22(3), September, 159-67 and in Maria Saleth (Eds), 2002; *Water Resource and Economic Development*, Edward Elgar Publishing Ltd., UK.

Joshi, Archana (2006): *Groundwater Conservation: A Perspective Plan for Ahmedabad Urban Complex Area, Master of Planning* Dissertation Report (unpublished), School of Planning, CEPT University, Ahmedabad.

Joshi, M.B. (2001): "Operation of Sardar Sarovar Conveyance System", *International Journal of Water Resources Development*, Vol. 17, 1 March.

— and Kapadia, Vivek (2010): "Sharing Water in the 21st Century: Rethinking the Rationale", *Irrigation and Drainage*, 59: 92-101.

Joshi, Vidyut (1987): *Submerging Villages – Problems and Prospects*, Ajanta Publications (India). Delhi.

— (1991): *Rehabilitation – A Promise to Keep (A Case Study of SSP)*, The Tax Publications. Ahmedabad.

— (1991): *Rehabilitation at Sardar Sarovar Narmada Project* in Pathak, Mahesh T. (ed.): *Sardar Sarovar Project: A Promise for Plenty*. Oxford and IBH Publishing. New Delhi.

Kalpavriksh (1995): "Cry Our Beloved Narmada," *The India Magazine*, New Delhi. February.

Kay, Melvyn; Tom Franks and Laurence Smith (1997): *Water: Economics, Management and Demand*. London: E&FN Spon, London.

Khosla, A.N. (1964): Report submitted to the Central Water Commission. New Delhi.

Kothari, A. and S. Singh (1988): *The Narmada Valley Project: A Critique*, Kalpavriksh. New Delhi.

Kothari, Smitu (1996): "Whose Nation? Displaced as Victims of Development," *Economic and Political Weekly*, 31 (24).

Krishiworld.com http://www.krishiworld.com/html/water_resources2. html (accessed on 28th April, 2010)

Krishna Jai, Chaturvedr R.S., Srivastava L.S. (1976): *Geological and Seismological Investigations of the Environs of Narmada Valley around Navagam Dam Site in Gujarat State.*

—(1989): "Dams and Seismicity.".

Krishna Kumar (1996): "Narmada Debate: A Test of Modernity," *Indian Express*, July, 11.

Kumar, M. Dinesh (2002): Reconciling Water Use and Environment: Water Resource Management in Gujarat, Resource, Problems, Issues, Strategies and Framework for Action, *Report prepared for State Environmental Action Project* supported by the World Bank. Gujarat Ecology Commission.

— (2005): "Impact of Electricity Prices and Volumetric Water Allocation on Energy and Groundwater Demand Management: Analysis from Western India", *Energy Policy*, 33 (1).

— (2007): *Groundwater Management in India: Physical, Institutional and Policy Alternatives*, New Delhi: Sage Publications.

— (2009): *Water Management in India: What Works, What Doesn't*, New Delhi: Gyan Books.

— (2010): *Managing Water in River Basins: Hydrology, Economics, and Institutions*, New Delhi: Oxford University Press.

— and Jos van Dam (2009): "Improving Water Productivity in Agriculture in India: Beyond 'More Crop per Drop'", in Kumar, M. Dinesh and Upali Amarasinghe (Eds) Water Productivity Improvements in Indian Agriculture: Potentials, Constraints and Prospects, *Strategic Analysis of National River Linking Project of India Series 4*. Colombo: International Water Management Institute.

— and O.P. Singh (2001): "Market Instruments for Demand Management in the Face of Scarcity and Overuse of Water in Gujarat," *Water Policy*, 3 (5).

— (2005): "Virtual Water in Global Food and Water Policy Making: Is There a Need for Rethinking?" *Water Resources Management*, 19: 759-789.

— and Upali Amarasinghe (Eds) (2009): "Water Productivity Improvements in Indian Agriculture: Potentials, Constraints and Prospects", *Strategic Analysis of National River Linking Project of India Series 4*. Colombo: International Water Management Institute, Colombo.

— A. Narayanamoorthy, O.P. Singh, M.V.K. Sivamohan, Manoj Sharma and Nitin Bassi (2010), "Gujarat's Agricultural Growth Story: Exploding Some Myths", *Occasional Paper 2*, Institute for Resource Analysis and Policy, Hyderabad.

— Ajaya Kumar Malla and Sushanta Tripathy (2008): "Economic Value of

Water in Agriculture: Comparative Analysis of a Water-Scarce and a Water-Rich Region in India", *Water International*, 33 (2): 214-230.

Kumar, M., A. Dinesh, A. Narayanamoorthy and M.V.K. Sivamohan (2010), Gujarat's *Agricultural Growth Story: Exploding Some Myths*. Occasional Paper 2, Institute for Resource Analysis and Policy, Hyderabad.

Kumar, M.D. (1995): "Dynamics of drinking water in rural Gujarat: physical, social and institutional dimensions", paper presented in the workshop of *Drinking Water Problems in Rural Gujarat: Issues and Perspectives*, GIDR, January 25.

Levenhagen, D.W. (1987): *Overview Environmental Impact Report: Sardar Sarovar and Narmada Sagar Complex Project Area, Narmada River Basin, India*. Consultant's Report. World Bank. Washington D.C.

Lobo, L., S. Kumar (2009): *Land Acquisition, Displacement and Resettlement in Gujarat,1947-2004,* SAGE Publications, New Delhi.

Lobo, Lancy and Shashikant Kumar (2009): *Land Acquisition, Displacement and Resettlement in Gujarat: 1947-2004*, Sage India, New Delhi.

Luders, Rolf J. (1991), "Massive Divestiture and Privatization: The Case of Chile," *Contemporary Policy Issues*, Vol. 9, No. 4, pp. 1-19.

M.S. University, Baroda (1993): *Studies on Ecology and Environment*, Department of Geology, Project Report submitted to SSNNL, Government of Gujarat.

Mahapatra, L.K. (1990): "Rehabilitation of Tribals Affected by Major Dams and Other Projects in Orissa," in A.P. Fernandez (ed.) *Workshop on Rehabilitation of Persons Displaced by Development Projects: A Report*, Institute for Social and Economic Change and MYRADA. Bangalore.

Mallik, Pradeep (2010): "Dams! Sarovar Project A Mess," *Ahmedabad Mirror*, Ahmedabad. March.

Manthan Adhyayan Kendra (2005): *Unravelling Bhakra*. (A Report), Badwani, Madhya Pradesh, India. April.

Marie-Helene, Z. (2006): *Urban Water and Waste Water*. New Delhi: Oxford University Press.

Mathew, A. (1989): "Industries, Mines and Disposition of Indigenous People," in W. Fernandes and Enakshi Ganguly Thukral (eds.) *Development, Displacement and Rehabilitation*, Indian Social Institute. New Delhi.

McKinsey Global Institute (2010): *India's Urban Awakening: Building Inclusive Cities, Sustaining Economic Growth*. April.

Meade, Ronald B. (1991): *Reservoirs and Earthquakes. Engineering Geology*. Volume 30, Issue 3-4.

Mehta, S.S. (2010): "Access to Urban Water and Sanitation to the Poor in Gujarat and Maharashtra". *A Research Report Prepared for PAS Project*, (draft, mimeo), CEPT University.

Meinzen-Dick Ruth S., L. Brown, H.S. Feldstein and A.R. Quisumbing (1997): "Gender, Property Rights and Natural Resources," in *World Development*, 25 (8), pp. 1303-1315.

Merrey, D.J. (1996): "Institutional Design Principles for Accountability in Large Irrigation Systems", *Research Report 8*. Colombo: International Water Management Institute.

MHUPA and UNDP (2009): *India: Urban Poverty Report 2009,* Oxford University Press, New Delhi.

Milgroms, Paul and Roberts, John (1992): *Economics, Organization & Management*, Prentice Hall Publication, pp 543.

Ministry of Agriculture and Cooperation (1985): *Report of the Inter-departmental Committee on Soil Conservation and Afforestation, Sardar Sarovar and Narmada Catchment.* Government of India. New Delhi.

Ministry of Environment & Forests (1986): *Environmental Aspects of Narmada Sagar and Sardar Sarovar* – A Study for Narmada Planning Group.

— (1986): *Cost Benefit Analysis.* Circular No.11-61/85-FRY (Cons). Government of India, January, 13.

— (1987a): Office Memorandum No. 3-87/80-1A. June, 24.

— (1987b): Memorandum No. 8-372/83-FC. September, 8.

— (1987c): Memorandum No 8-646//84-FC. October, 7.

— (1990): Memorandum No.8- 29/89-FC. March, 15.

Ministry of Irrigation and Power (1972): *Report of the Irrigation Commission.* Government of India. Volume III, Part I. New Delhi.

Ministry of Urban Development (2008): *Handbook of Service Level Benchmarks*, Government of India.

Mitra, Amit (1995): "Development Distress," *Down to Earth*, Vol. 3, No. 21.

Modi, N. (2011): *Convenient Action: Gujarat's Responses to Challenges of Climate Change*, Macmillan Publishers India, New Delhi.

Morse, Bradford and Thomas Berger (1992): *Sardar Sarovar: Report of the Independent Review*, Resource Futures International Inc. Ottawa.

Musgrave, Richard A. and Peggy B. Musgrave (1989): *Public Finance in Theory and Practice*, 5th Edition. McGraw-Hill Book Company. New York.

Muthayya, B.C. and J.S. Mathur (1975): *Rehabilitation of Displaced Villagers*, National Institute of Community Development. Hyerabad.

Nagi Reddy, C.P. *et al.* (1981): *Relocation and Resettlement Plan for Project Affected Persons Under Singoor Project*, Centre for Economic and Social Studies. Hyderabad.

Naik, P.J., R.S. Patil and R.S. Maradi (1990): "Reservoir Submergence and Rehabilitation in Six Villages in the Upper Krishna Project," in A.P.

Fernandez (ed.) *Workshop on Rehabilitation of Persons Displaced by Development Projects: A Report*, Institute for Social and Economic Change and MYRADA. Bangalore.

Nankani, Helen B. (1990): "Lessons in Privatization in Developing Countries," *Finance and Development*, Vol. 27, No. 1, pp. 43-45.

Narmada Bachao Andolan (1990): *Narmada – A Campaign Newsletter*, New Delhi. December.

— (1992): "Response to ARCH-Vahini's Critique of Narmada Bachao Andolan and Anti-dam Arguments," *(Mimeo)*.

— (1990): *Sardar Sarovar Project: An Economic, Environmental and Human Disaster*. Bombay.

Narmada Control Authority (1991): *Minutes of Meeting: Review of Activities of the Environmental Subgroup*, Item No XXXVIII 6 (417).

— (1992): *Silting of Sardar Sarovar.* Publication Number 2/92

—(1993): *Seismicity and Sardar Sarovar Project.*

—(2007): *Status Report on Sardar Sarovar Project.*

Narmada Planning Group (1981): *Projectised Studies: Sardar Sarovar (Narmada) Project.* Government of Gujarat.

—(1989): *Planning for Prosperity: Sardar Sarovar Development Plan.* Gandhinagar.

—(1996): *Report of the Expert Committee on financial Resources for Sardar Sarovar Project.*

Narmada Control Board (1989): *Minutes of Meeting: Rehabilitation Subgroup*, Narmada Valley Development Authority. December.

Narmada Planning Group (n.d.): "Cropping pattern and Water demand study in Narmada Command Area", Volume-I *Base Studies*, Government of Gujarat, Gandhinagar.

Narmada Project Dam Design Circle (1982 and 1983): *Sedimentation Studies and Life of Reservoir: Sardar Sarovar (Narmada) Project (FRL 455 ft).* Vadodara.

Narmada Valley Development Authority (1992): *Action Plan of Rehabilitation and Resettlement of Oustees of Madhya Pradesh (Sardar Sarovar Project)*, Parts I and II. Government of Madhya Pradesh. Bhopal. January.

— (1993): *The Bhil Track: A Study of Displaced Tribals*, Government of Madhya Pradesh. Bhopal.

Narmada Water Dispute Tribunal (1978): *Report of the Narmada Water Dispute Tribunal.* Government of India. New Delhi.

—(1978): *Further Report on the Narmada Water Dispute Tribunal with it's Decision.* Government of India. New Delhi.

Narmada Water Disputes Tribunal Award (1979): *Sub-Clause IV(1) – Provision for Rehabilitation*, December.

Narmada Water Disputes Tribunal. (1979): *The Report of the Narmada Water Disputes Tribunal,* Volumes 1 and 2. New Delhi.

National Commission on Farmers (2005): *Serving Farmers and Saving Farming – Jai Kisan: A Draft National Policy for Farmers*, Draft Fourth Report, Ministry of Agriculture, Government of India.

National Remote Sensing Centre (2009): *Assessment of Irrigation Potential created in ABIP funded Irrigation Projects in India using Cartosat Satellite data*, Hyederabad.

Nitin Desai Expert Committee (1983): *Report of the Committee to Review the Existing Criteria for Working out Benefit Cost Ratio for Irrigation Projects.* Government of India.

Padhiari, H., (2005): "Water Service Markets in Surface Irrigation Systems: Institutions and Socio-Economic Impacts." IWMI-Tata Water Policy Programme, paper presented in *3rd ITP Annual Partners' Meet*, (unpublished research) Anand.

Paranjpye, V. (1989): *The Narmada Valley Project: A Holistic Evaluation of the Sardar Sarovar and Indira (Narmada) Sagar Dams,* Indian National Trust for Art and Cultural Heritage (INTACH). New Delhi.

— (1990): *High Dams on the Narmada*, INTACH. New Delhi.

Paranjpye, Vijay (1990): *High Dams on the Narmada: A Holistic Analysis of the River Valley Projects*, Indian National Trust for Art and Cultural Heritage. New Delhi.

Parasuraman, S. (1993): "The Anti-Narmada Movement in India: Can the Resettlement and Rehabilitation Policy Gains be Translated into a national Policy," *Working Paper Series, No.161*, Institute of Social Studies, The Hague.

— (1996): "Outcomes for Displaced: Two Case Studies," *Economic and Political Weekly*, 31 (24).

Parthasarathy, R. (2000): "Participatory irrigation Management Programme in Gujarat: Institutional and Financial Issues," *Economic and Political Weekly*, Vol. 35 (35 and 36), August, 26 to September 2.

— (2002): "Reforms in Irrigation Management and Peoples' Participation," in Rakesh Hooja, Ganesh Pangare and K.V. Raju (ed.) *Users in Water Management: The Andhra Model and its Replicability in India,* Rawat Publications, Jaipur and New Delhi.

— (2008): "Irrigation Water Pricing: Analytic of Competing Sources," in Vishwa Ballabh (ed.) *Governance of Water: Institutional Alternatives and Political Economy,* Sage, New Delhi.

— and Jharna Pathak (2008a): "The Guiding Visible Hand of Participatory Approaches to Irrigation Management," in V Ratna Reddy and S Mahendra Dev (ed.) *Managing Water Resources: Policies, Institutions, and Technologies,* Oxford University Press, New Delhi.

— and P.K. Vishwanathan (2008b): "Are Water Policies A Case of Reverse Engineering in India?" *Proceedings of the 7th Annual Partners Meet Volume - 2*, IWMI-TATA Water Policy Research Programme, Hyderabad, April.

— (2010): "The Role of Irrigation in the Growth Story of Gujarat," in Ravindra Dholakia and Samar K. Datta (ed.) *High Growth Trajectory and Structural Changes in Gujarat Agriculture*, Macmillan Publishers, New Delhi.

Patel, A.S. and H.F. Patel (1992): *Some Policy Issues for Canal Water Irrigation in Gujarat.* Department of Economics, Sardar Patel University.

Patel, Anil (2001), "Resettlement in the Sardar Sarovar Project: A Cause Vitiated". *International Journal of Water Resources Development*, 17 (3): 315-328.

— *et al.* (1990): "The Tragic Story of the Displaced: A Case Study of the Tribals Affected by Sardar Sarovar," in A.P. Fernandez (ed.) *Workshop on Rehabilitation of Persons Displaced by Development Projects: A Report*, Institute for Social and Economic Change and MYRADA. Bangalore.

— (1991): "Rehabilitation and Resettlement in Sardar Sarovar Project: Are the Critics Right?" *Social Action.* New Delhi. July-September.

— (1994): "What do the Tribals Want in the Valley of Narmada?" Paper presented at an *International Conference on Indigenous People and Water Reservoir Development Projects*, Montreal, Canada, April.

— (1995): "Reforms or Derailment: Activists' Dilemma Revisited," Paper presented at Refugee Studies Programme (RSP) Conference on *Development induced Displacement and Improvement*, Oxford, U.K. January.

— (1997): "What do the Narmada Valley Tribals Want?" in W.F. Fisher (ed.) *Towards Sustainable Development*, Rawat Publications. Jaipur.

— (2001): "Resettlement in Sardar Sarovar Project," *Water Resource Development*, Vol. 7(3).

—; A. Mehta; S. Iyengar; T. Parekh and R. Mishra (1992): *Morse-Berger Review: A Search for Truth,* ARCH-Vahini. Mangrol.

Patel, H.G. (1990): *Financial Working of Irrigation Projects—A Case Study of Four Projects in Gujarat.* Department of Economics, Sardar Patel University.

Patel, V.M. (1991): "Sardar Sarovar Project : Benefit-Cost Analysis" in M.T. Pathak (ed.) *Sardar Sarovar Project: A Promise for Plenty*, Oxford and IBH Publishing Co. Pvt. Ltd., New Delhi.

Pathak, M., V.D. Shah (2010): "Five Decades of Gujarat Agriculture: Some Reflections" in R.K. Dholakia, S.K. Datta (Eds): *High Growth Trajectory and Structural Changes in Gujarat Agriculture*. Macmillan Publishers India Ltd.

Pathak, Mahesh (1989): *Environmental Planning for Sardar Sarovar project.* SSNNL. Gandhinagar.

—(ed.) (1991): *Sardar Sarovar Project: A Promise for Plenty.* Oxford and IBH Publishing. New Delhi.

—(ed.) (1995): *Development Imperatives and Environmental Concerns, Sardar Sarovar Project.* Agro Economic Research Centre. Vallabh Vidyanagar.

Patkar, Medha (1988): *Sardar Sarovar: Vikas Ki Vinash* (in Hindi), Parisar, Pune.

Perry, Chris J. (2001a): "World Commission on Dams: Implications for Food and Irrigation", *Irrigation and Drainage*, 50: 101-107.

— (2001b), Water at any Price? Issues and Options in Charging for Irrigation Water, *Irrigation and Drainage*, 50 (1): 1-7.

Plusquellec, H. (2002): *How design, management and policy affect the performance of irrigation projects: Emerging modernisation procedures and design standards*. Food and Agriculture Organisation of the United Nations, Bangkok ftp://ftp.fao.org/docrep/fao/004/ac799e/ac799e00.pdf (accessed on 3rd February, 2010) 20.

Raj, P. A. (1990): *Facts: Sardar Sarovar Project*, Sardar Sarovar Narmada Nigam, Gandhinagar.

Raj, P.A. (1991): Erroneous Zones, *All About Narmada*, Government of Gujarat, Gandhinagar.

Ram, Rahul (1993): *Muddy Waters,* Kalpavriksh. New Delhi.

Ranade, Rahul and M. Dinesh Kumar (2004), "Narmada Water for Groundwater Recharge in North Gujarat: Conjunctive Management in Large Irrigation Projects", *Economic and Political Weekly*, XXXIX (31): pp. 3498-3503.

Rao, S.V.L.N. (1992): *Pricing of a Public Utility: A Survey.* Department of Economics, South Gujarat University.

Raskin, P., P. Gleick, P. Kirshen, G. Pontius and K. Strzepek (1997): "Water Futures: Assessment of Long-Range Patterns and Problems", Stockholm Environment Institute, Stockholm.

Ravindra H. Dholakia and Samar K. Dutta (Eds) (2010): *High Growth Trajectory and Structural Changes in Gujarat Agriculture*, Macmillan Publishers, New Delhi.

Reddy, I.U.B. (1992): *Displacement and Rehabilitation*, Mittal Publications. New Delhi.

— (1994): *Industrial Development and Problems of the Uprooted*, Rawat Publications. Jaipur.

Rich, Bruce (1988): *Statement before the Subcommittee on Foreign Operations Committee on Appropriations*, US Senate. June, 14.

Rondinelli, Dennis A. and John D. Kasarda (1991): Privatizing Public Services in Developing Countries: What Do We Know?", *Business & the Contemporary World*, Vol. 3, No. 2, Winter, pp. 102-113

Rosegrant, Mark W. and Hans P. Binswanger (1994): "Markets in Tradable Water Rights: Potential for Efficiency Gains in Developing Country Water Resource Allocation", *World Development*, 22 (11): pp. 1613-1625.

— and Renato Gazmuri Schleyer (1994), "Reforming Water Allocation Policy through Markets in Tradable Water Rights: Lessons from Chile, Mexico, and California", EPTD Discussion paper 2, International Food Policy Research Institute, Washington.

— (1996) Establishing Tradable Water Rights: Implementation of the Mexican Water Law, *Irrigation and Drainage Systems*, 00: 263-279.

RTI India (2008): Gujarat used only 7% Narmada dam water. As reported in Yahoo News 29th May, 2008. http://www.rtiindia.org/forum/4391-gujarat-used-only-7-narmada-dam-water.html (accessed on 3rd January, 2010).

Sah, D.C. (1993): *Monitoring and Evaluation of Resettlement and Rehabilitation Programme for Sardar Sarovar (Narmada) Project*, Report 16. Centre for Social Studies. Surat.

— (1994a): *Monitoring and Evaluation of Resettlement and Rehabilitation Programme for Sardar Sarovar (Narmada) Project*, Report 19. Centre for Social Studies. Surat.

— (1994b): "Avoiding Impoverishment Risks: Towards a Policy Frame of Dam Affected People, " Paper presented for the Seminar on *Development, Displacement and Rehabilitation* at Centre for Social Studies. Surat. 11-13 May. Revised version of the paper was accepted for publication in *Review of Development and Change* [Vol. 3 (1) 1998].

— (1994c): *Commercialisation in Agriculture and Living Condition,* (M&E Report 21), Centre for Social Studies. Surat.

— (1995a): *Monitoring and Evaluation of Resettlement and Rehabilitation Programme for Sardar Sarovar (Narmada) Project*, Report 20. Centre for Social Studies. Surat.

— (1995b): "Development and Displacement: National Rehabilitation Policy." *Economic and Political Weekly*, Vol. 30 (48). December 2.

— (1996a): *Monitoring and Evaluation of Resettlement and Rehabilitation Programme for Sardar Sarovar (Narmada) Project*, Report 21. Centre for Social Studies. Surat.

— (1996b): *Monitoring and Evaluation of Resettlement and Rehabilitation Programme for Sardar Sarovar (Narmada) Project*, Draft Report 23. Centre for Social Studies. Surat.

— (1996c): "Resettlement and Rehabilitation: Sardar Sarovar Project on the Narmada" *(Mimeo),* Report No. 22. Centre for Social Studies. Surat.

— (1996d): *Living Condition in 31 New Sites* (M & E Report 24), Centre for Social Studies. Surat.

— (1997): "Second Generation Problems of Involuntary Migration," *Indian Journal of Agricultural Economic,* Vol. 52(4).

— (1998): "Avoiding Impoverishment Risks: Towards a Policy Frame of Dam Affected People," *Review of Development and Change*, Vol. 3(1).

— (1999): "Pressures on Land, Employment and Migrations," *Review of Development and Change,* Vol. 4(2).

— (2003): *Involuntary Migration: Evidence from Sardar Sarovar Project,* Rawat Publications, Jaipur.

— and J.N. Soni (1993): *Monitoring and Evaluation of Resettlement and Rehabilitation Programme for Sardar Sarovar (Narmada) Project*, Report 15. Centre for Social Studies. Surat.

Saleth, R.M. (1999): "Irrigation Privatization in India: Options, Framework and Experience", *Economic and Political Weekly*, 34(26): A86-A92.

Sardar Sarovar Narmada Nigam Limited (SSNNL) (1988): *Planning for Prosperity. Sardar Sarovar Development Plan*, Narmada Planning Group, Gandhinagar.

Scudder, Thayer (1991): "A Sociological Framework for the Analysis of New Land Settlements" in Michael M. Cernea (ed.) *Putting People First: Sociological Variables in Rural Development*, Oxford University Press, Second Edition. New York.

— (2003): *India's Sardar Sarovar Project,* (Unpublished Manuscript).

Sen, A.K. (1981): *Poverty and Famines,* Oxford University Press. Oxford.

Shah, G. and Biswaroop Das (1986): "Voluntary Organisation Development," *(Mimeo).* Centre for Social Studies. Surat.

— (1988): "Voluntary Organisation in Gujarat," *(Mimeo).* Centre for Social Studies. Surat.

Shah, H. (2009): *Narmada Yojana: Nehro ne badle pipeline ketli vyajabi?* (Gujarati) Arthbodh.

Shah, T. (1993): "Groundwater Markets and Irrigation Development", *Political Economy and Practical Policy*. Oxford University Press, Bombay.

— (2009): *Taming the Anarchy. Groundwater Governance in South Asia*, Washington DC: RFF Press.

—, S. Bhatt, R.K. Shah, J. Talati (2008): "Groundwater Governance through Electricity Supply Management: Assessing an Innovative Intervention in Gujarat, Western India". *Agriculture Water Managegment,* 95(11), 1233-1242.

—, S. Bhattacharya (1993): *Farmer Organisations for Lift Irrigation: Irrigation Companies and Tubewell Cooperatives of Gujarat*, Network Paper 26. Overseas Development Institute, London.

—, S. Verma (2008): "Co-management of Electricity and Groundwater: An Assessment of Gujarat's Jyotirgram Scheme". *Economic and Political Weekly,* Vol.43(7), 59-66.

Shah, T., A. Gulati, P. Hemant, G. Shreedhar, R.C. Jain (2009): "Secret of Gujarat's Agrarian Miracle after 2000". Review of Agriculture. *Economic and Political Weekly*, Vol. XlIV, no. 52.

Shah, Tushaar (1991): "Water Markets and Irrigation Development in India," *Indian Journal of Agricultural Economics*, Vol. 46, No. 3, July-September.

— (1993): *"Ground Water Markets and Irrigation Development in India: Political Economy and Practical Policy,"* Oxford University Press. Bombay.

Shah, V.I., Desai, N.V. and S.V. Joshi (1997): *Control of Brackishness in Drinking Water*, Central Salt and Marine Chemicals Research Institute (CSMCRI), Bhavnagar.

Shah, Zankhana and M. Dinesh Kumar (2008), "In the Midst of the Large Dam Controversy: Objectives, Criteria for Assessing Large Water Storages in the Developing World", *Water Resources Management*, 22 (12): 1799-1824.

Sharan, Girja (2007): *Harvesting Dew to Supplement Drinking Water Supply in Arid Coastal Villages of Gujarat, Paper presented in workshop on "Crisis in Drinking Water in Coastal India*, EDI, Ahmedabad.

Sharma, U.S. (1996): "Some Aspects of Drinking Water Supply in Rural Gujarat (Adequacy and Operation – Maintenance)", Proceedings of National Seminar On *Rural Water Supply Sanitation*, Organised by Centre For Development Studies, Thiruvananthapuram.

Sharma, Usha (1995): *A Study on Some Aspects of Rural Drinking Water Supply in Gujarat: Adequacy, Quality and Operation-Maintenance*, Sardar Patel Institute of Economic and Social Research, Ahmedabad.

Sheth, Pravin (1991): *Sardar Sarovar Project: Dynamics of Development*, Vikas Bharati Institute of Policy Studies, Research and Futurology. Ahmedabad.

— (1995): "Politics of Narmada Project," *(Mimeo)*. Madhya Pradesh Institute of Social Science Research. Ujjain.

Shingi, Prakash M. and V.N. Asopa (2002): *Independent Evaluation of Check dams in Gujarat Strategies and Impacts*, Centre for Management in Agriculture, Indian Institute of Management, Ahmedabad.

Shroff, Manu (1995): *Economic Viability of SSP* in Pathak, Mahesh (ed.) *Development Imperatives and Environmental Concerns.* Agro Economic Research Centre. Vallabh Vidyanagar.

Shunglu, V.K., G.K. Chadha and Jaiprakash Narayan (2006): *Report on the Status of Rehabilitation of Project Affected Families in Madhya Pradesh*, Sardar Sarovar Project Relief & Rehabilitation Oversight Group.

Singh, O.P. (2004): "Water-intensity of North Gujarat's Dairy Industry: Why Dairy Industry Should Take a Serious Look at Irrigation?", Paper presented at the Second International Conference of the *Asia Pacific Association of Hydrology and Water Resources* (APHW 2004), Singapore, June 5-9.

Singh S, (2003): "Large Dams and Environment" in Prasad K (ed.) *Water Resource and Sustainable Development*, Shipra Publications, Delhi.

— (2006): "Credit, Indebtedness and Farmer Suicides in Punjab: Some Missing Links." *Economic and Political Weekly*. 41(3), 3330–31.

Singhal, N., V. Patwari, V. (2009): "Evolving arrangements for local water diversion-delivery in SSP." *Management Traineeship Segment Report*, Institute of Rural Management, Anand.

Sinha, B.K. (1996): "National Policy for Rehabilitation: Objectives and Principles," *Economic and Political Weekly*, 31 (24).

Sneh Setu Foundation (2008): *Impact of Sardar Sarovar Project Water on Selected Villages of Gujarat*, A study report, September, Vadodara.

Solanki, Rupa (1994): *Condition of Tribal Women in Relocation Sites*, A report submitted towards partial fulfilment of coursework in Master of Social Work, Gujarat Vidyapith. Ahmedabad.

Soni, Trupti and Gautam Appa. (1993): "Sardar Sarovar Project Downstream Impact," Paper presented in *Workshop at Narmada Forum*, Delhi School of Economics and Institute of Economic Growth.

SSNL (1988): *Planning for Prosperity. Sardar Sarovar Development Plan*, Narmada Planning Group, Gandhinagar.

— (2003): *Environment & Amelioration Programme (Fisheries)*, Gandhinagar, January. No. 1," in *Texas Canal Lining and Pipeline Installation Project*", http://www.cocef.org/aproyectos/ExComHidalgo2003_05ing.htm (accessed on 28th December, 2009)

— (2009): "Benefits-Irrigation", Sardar Sarovar Narmada Nigam Limited. <http://www.sardarsarovardam.org/benefits.htm> (accessed on 12 October, 2009).

— (2009): *Water Use Plan—SSP.* Gandhinagar.

—(2010): *Report of the Expert Group for Strategy for Accelerated Development of Sardar Sarovar Project command Area.* Gandhinagar.

— (2010): Personal discussions with Executive Director, Agriculture, Sardar Sarovar Narmada Nigam Limited, Gandhinagar.

SSNNL's letter of 24/04/98 to Executive Member, NCA.

Sudarsen, V. and M.A. Kalam (eds.) (1990): *The Uprooted: Development, Displacement and Resettlement,* Gian Publishing House. Delhi.

Supplementary Affidavit to the Writ Petition (Civil) 319. 319 (Supreme Court of India, July 1997).

Supreme Court of India (2002): *Arundhati Roy.... Contemner, Contempt Petition (crl.) 10 of 2001*, Judgment, March 6, 2002, New Delhi.

SWRC (1988): "Towards People's Power in Jhabua," *People's Action.* May.

TAHAL Consulting Engineers Limited, Israel (1987): *Water Resources Planning for the State of Gujarat,* A report submitted to SSNNL, Gandhinagar, Gujarat.

Talati, J., J. Liebrand (2003): "Evolving institutions for irrigation management in Sardar Sarovar Project Command", *IWMI-Tata Water Policy Program*,(unpublished research) Anand.

—, T. Shah (2004): "Institutional Vacuum in Sardar Sarovar Project: Framing 'Rules-of-the-Game'". *Economic and Political Weekly*, 39(31), 3504–509.

Tata Economic Consultancy Services (1981): *Methodological Framework for Economic Appraisal of Narmada Project.*

Tata Institute of Social Sciences (1993): *The Sardar Sarovar Project: Experiences with Resettlement and Rehabilitation*, Tata Institute of Social Sciences. Bombay.

Thakkar, Himanshu (2005): "Who Pays for Our Giant Follies?" *Info-change News & Features,* October.

The Hindu (2006): *GoM's confidential report on A Brief Note on the Assessment of Resettlement and Rehabilitation (R&R) Sites and Submergence of Villages of the Sardar Sarovar Project.* Note of the Group of Ministers Dated 9 April 2006. Chennai. April 17.

Thobani, Mateen (1997), "Formal Water Markets: Why, When and How to Introduce Tradable Water Rights", *The World Bank Research Observer*, 12 (2): 161-179.

Thukral, Enakshi Ganguly (1989): "Dams for Whose Development?" in W. Fernandes and E. Ganguly Thukral (eds.) *Development, Displacement and Rehabilitation: Issues for a National Debate*, Indian Social Institute. New Delhi.

— (ed.) (1992): *Big Dams, Displaced People: Rivers of Sorrow Rivers of Change,* Sage Publications. New Delhi.

Times of India (2009a): "Shelve Narmada pipeline plan: Expert." Ahmedabad edition, September 12.

— (2009b): "Narmada pipeline project scaled down." Ahmedabad edition, November 13.

Tolba, M.K. (1988): *Sustainable Water Development: Opportunities and Constraints*, Vol. 13 (4).

TRENDSNIFF. (2008, September): Retrieved December 2010, from www.trendsniff.com: http://trendsniff.com/2008/09/16/targeting-top-20-growth-cities-in-india/

United States Department of the Interior, Bureau of Reclamation (1976): *Design of Gravity Dams.* Denver.

— (1983): *Economic Appraisal of Sardar Sarovar Project.* Bombay.

Unknown (2009): "State to Press for a Taller Narmada Dam," *Indian Express.* October 23.

US Bureau of Reclamation (1973): *Water System Automation*, Current Information on Automation of Reclamation Water Systems, United States Department of the Interior, Denver, Colorado.

Vaidyanathan, A. (2006): *India's Water Resources – Contemporary Issues on Irrigation*, Oxford University Press, New Delhi.

Varghese, B.G. (1994): *Winning the Future: From Bhakra to Narmada, Tehri, Rajasthan Canal*, Konark Publishers Pvt. Ltd., New Delhi.

Varma, S.C. (1985): *Human Settlement in Lower Narmada Basin*, Narmada Valley Development Authority. Bhopal.

Vermillion, Douglas (1997): *Impacts of Irrigation Management Transfer*, IIMI, Research Report 11, April.

Vice Chairman, NVDA's letter No. NVDA/D (Fish)/88/3867 of 1/11/88 addressed to the Chairman, NCA.

Vyas, J. (2001), "Water and energy for development in Gujarat with special focus on the Sardar Sarovar project", *International Journal Water Resources Development*, 17(1):37–54.

Vyas, Jay Narayan (2001): "Water and Energy for Development in Gujarat with Special Focus on the Sardar Sarovar Project," *International Journal of Water Resources Development,* Vol. 17, 1, March.

Wallingford, H.R. (1993): *Environmental Changes Downstream of Sardar Sarovar Dam.* Ex 2750.

—(1993): *Sardar Sarovar Project Hydrology and Water Resources.*

—(1993): *Sediment and Backwater Aspects of SSP Report.* Ex 2761.

—(1993): *SSP: Command Area Environmental Impact Assessment Progress Report.*

—(1995): *Potential for Agro Chemical Pollution from the Sardar Sarovar Command Area, India.* Ex 3177.

Wallingford, H.R. (1995): *Environmental Changes Downstream of Sardar Sarovar*. Oxfordshire, UK. Report No. EX 2750. of World Bank.

Wamana Consultants (1994): *Study of Preparation of Detailed Command Area Development Plan: SSP.*

2030 Water Resources Group (2009): *Charting Our Water Future: Economic Frameworks to Inform Decision-Making*, 2030WaterResourcesGroup @mckinsey.com

Watt, Jacqueline (2008), "The Effect of Irrigation on Surface-Ground Water Interactions: Quantifying Time Dependent Spatial Dynamics in Irrigation Systems," Thesis submitted to Charles Sturt University for the Degree of Doctor of Philosophy, School of Environmental Sciences, Faculty of Sciences, Charles Sturt University, March.

Wood, John R. (1993): "India's Narmada River Dam: Sardar Sarovar under Siege," *Asian Survey*, Vol. 33, No. 10.

World Bank (1985): *Narmada River Development, Gujarat—Sardar Sarovar Dam and Power Project*, Staff Appraisal Report 5107-IN; and *Water Delivery and Drainage Project*, Staff Appraisal Report 5108-IN.

—(1990): *Review of Sardar Sarovar (Narmada) Projects Economic Analysis*. India Agriculture Operations Division.

—(1993): *Project Completion Report: Narmada River Development, Gujarat.*

—(1995): *Project Completion Report-India-Narmada River Development-Gujarat Sardar Sarovar Dam and Power Project*, Report No. 14159.

— (1995): *Project Completion Report (Sardar Sarovar).* World Bank.

— (2005): *India's Water Economy: Bracing for a Turbulent Future*, World Bank, Washington.

Zoology Department (1994): *Environmental Impact Assessment Studies on inland and Marine Fisheries Relevant to the Command Area of Sardar Sarovar Project.* M.S. University. Baroda.

Index

Consolidated Index of all 3 Vols.